RILEM State-of-the-Art Reports

RILEM STATE-OF-THE-ART REPORTS
Volume 34

RILEM, The International Union of Laboratories and Experts in Construction Materials, Systems and Structures, founded in 1947, is a non-governmental scientific association whose goal is to contribute to progress in the construction sciences, techniques and industries, essentially by means of the communication it fosters between research and practice. RILEM's focus is on construction materials and their use in building and civil engineering structures, covering all phases of the building process from manufacture to use and recycling of materials. More information on RILEM and its previous publications can be found on www.RILEM.net.

The RILEM State-of-the-Art Reports (STAR) are produced by the Technical Committees. They represent one of the most important outputs that RILEM generates – high level scientific and engineering reports that provide cutting edge knowledge in a given field. The work of the TCs is one of RILEM's key functions.

Members of a TC are experts in their field and give their time freely to share their expertise. As a result, the broader scientific community benefits greatly from RILEM's activities.

RILEM's stated objective is to disseminate this information as widely as possible to the scientific community. RILEM therefore considers the STAR reports of its TCs as of highest importance, and encourages their publication whenever possible.

The information in this and similar reports is mostly pre-normative in the sense that it provides the underlying scientific fundamentals on which standards and codes of practice are based. Without such a solid scientific basis, construction practice will be less than efficient or economical.

It is RILEM's hope that this information will be of wide use to the scientific community.

Indexed in SCOPUS, Google Scholar and SpringerLink.

More information about this series at http://www.springer.com/series/8780

Aitor Llano-Torre · Pedro Serna

Editors

Round-Robin Test on Creep Behaviour in Cracked Sections of FRC: Experimental Program, Results and Database Analysis

State-of-the-Art Report of the RILEM TC 261-CCF

Editors
Aitor Llano-Torre ⓘ
Institute of Concrete Science
and Technology ICITECH
Universitat Politècnica de València
Valencia, Spain

Pedro Serna ⓘ
Institute of Concrete Science
and Technology ICITECH
Universitat Politècnica de València
Valencia, Spain

ISSN 2213-204X   ISSN 2213-2031   (electronic)
RILEM State-of-the-Art Reports
ISBN 978-3-030-72738-3   ISBN 978-3-030-72736-9   (eBook)
https://doi.org/10.1007/978-3-030-72736-9

This Springer imprint is published by the registered company Springer Nature Switzerland AG
The registered company address is: Gewerbestrasse 11, 6330 Cham, Switzerland

RILEM Technical Committee 261-CCF (Creep behaviour in Cracked Sections of Fibre Reinforced Concrete) Members

Chairman

Pedro Serna, Universitat Politècnica de València, Spain

Secretary

Sergio H. Pialarissi Cavalaro, Loughborough University, UK

Members

Bryan Barragan, Owens Corning, France
E. Stefan Bernard, Technologies in Structural Engineering Pty Ltd., Australia
William Peter Boshoff, University of Pretoria, South Africa
Nicola Buratti, University of Bologna, Italy
Todd Clarke, BarChip Inc., Australia
Vinh Dao, University of Queensland, Australia
Clementina del Prete, University of Bologna, Italy
Marco di Prisco, Politenico di Milano, Italy
Emilio Garcia-Taengua, University of Leeds, UK
Ravindra Gettu, Indian Institute of Technology Madras, India
Catherine Larive, CETU, France
Aitor Llano-Torre, Universitat Politècnica de València, Spain
Claudio Mazzotti, University of Bologna, Italy
Sandro Moro, Master Builders Solutions, Italy
Tomoya Nishiwaki, Tohoku University, Japan
Benoit Parmentier, CSTC-WTCB, Belgium

Preface

Fibre-reinforced concrete (FRC) is a special concrete with such a long enough background. Nevertheless, the introduction of FRC as a possible structural concrete in codes was relatively late, and thus, the application of FRC in real constructions was not as fast as expected. Among the aspects hindering the use in structures, the long-term behaviour or creep, mainly in the cracked state, is one of the main aspects that causes more reticence from the point of view of both the experimental characterization of creep and the incorporation in the calculation criteria. This fact was revealed in BEFIB2012 where many researchers presented their works on this topic with such different methodology proposals due to the lack of a standardized methodology. The RILEM Technical Committee 261-CCF was established in 2014 to lead with this new challenge.

One of the main objectives of the RILEM TC 261-CCF was to coordinate research joint efforts of the scientific community working on the topic and make possible the comparison of different creep test results. Due to significant differences in the proposed testing methodologies, the comparison of results was not reliable at all, even for similar methods. In this context, an international round-robin test (RRT) in the long-term behaviour of cracked FRC was proposed as an analytic tool to assess differences between available methodologies and find a unified procedure. This objective is required for the participants to uncover and expose their facilities, equipment and procedures so that the most significant differences between methodologies could be detected in the results analysis.

This book is the outcome of the work activities of the TC related to the RRT, and it was structured to expose step by step the work done in the RRT from the specimen's production to the creep tests results analysis and hence understand the obtained conclusions. Although the earliest idea was to write a monograph book, sometimes it becomes difficult to distinguish the authorship of the contributions for each chapter. I am sure that the author's list of each chapter is in fact more transversal than shown in the document. I would like to acknowledge the contribution of all TC members and external attendees during both face-to-face and online meetings, their proposals of ideas for the RRT experimental programme, their authorship contribution in the book and review of the many drafts of this manuscript. Details of TC members and affiliations are given after this Preface.

I also truly appreciate the contribution of all the people involved in the participant laboratories and institutions. Either researchers that accepted this challenge or technicians that performed the tests contributed to the success of the RRT. Their descriptions about equipment and procedures as well as their support to review all the data exchange improved the results interpretation and its ulterior analysis. Thanks for your valuable contribution.

Finally, special thanks go to Aitor Llano-Torre who has realized a huge work during last years to successfully conclude this RRT. He supervised the production and shipment of FRC specimens, managed the database exchange, compiled all the information, assessed the received data and coordinated the scientific analysis as well as the editorial work. The accomplishment of this work would not have been possible without his collaboration.

Valencia, Spain   Pedro Serna
                                                                    Chairman of the RILEM TC 261-CCF

Acknowledgements

This round-robin test (RRT) has been supported by the fibre suppliers Master Builders Solutions (BASF), BEKAERT and ArcelorMittal Fibres. The RRT organizers wish to thank their financial support in the fabrication and delivery of specimens all over the world.

The RRT organizers gratefully thank to the following institutions for their participation on this international experimental programme: Universitat Politècnica de València UPV, Universitat Politècnica de Catalunya UPC, OTH Regensburg, Master Builders Solutions (BASF), LEMIT-CIC and Facultad de Ingeniería UNLP, Indian Institute of Technology Madras (IITM), NV BEKAERT SA, Tohoku University, ArcelorMittal Fibres, BBRI Belgium Building Research Institute, University of Bologna, Universidade Federal de Rio de Janeiro, École Polytechnique de Montréal, Politecnico di Milano, Sigma Béton, CETU Centre for Tunnel Studies, Ministère de l'Ecologie, du Développement durable et de l'Energie (MEDDE), Stellenbosch University, TSE Technologies in Structural Engineering and VSH VersuchsStollen Hagerbach.

Besides the contribution of the authors, the contribution and suggestions of the RILEM TC 261-CCF official members B. Barragán, T. Clarke, M. di Prisco, G. Plizzari, K. A. Rieder, P. Rossi, D. Rogat, L. Vandewalle, G. Vitt and S. Wolf are greatly acknowledged as well as the valuable contribution to this report of the experts S. Pouillon, S. Jose, J. Bokern and J. R. Martí-Vargas.

RILEM Publications

The following list is presenting the global offer of RILEM Publications, sorted by series. Each publication is available in printed version and/or in online version.

RILEM Proceedings (PRO)

PRO 1: Durability of High Performance Concrete (ISBN: 2-912143-03-9; e-ISBN: 2-351580-12-5; e-ISBN: 2351580125); *Ed. H. Sommer*

PRO 2: Chloride Penetration into Concrete (ISBN: 2-912143-00-04; e-ISBN: 2912143454); *Eds. L.-O. Nilsson and J.-P. Ollivier*

PRO 3: Evaluation and Strengthening of Existing Masonry Structures (ISBN: 2-912143-02-0; e-ISBN: 2351580141); *Eds. L. Binda and C. Modena*

PRO 4: Concrete: From Material to Structure (ISBN: 2-912143-04-7; e-ISBN: 2351580206); *Eds. J.-P. Bournazel and Y. Malier*

PRO 5: The Role of Admixtures in High Performance Concrete (ISBN: 2-912143-05-5; e-ISBN: 2351580214); *Eds. J. G. Cabrera and R. Rivera-Villarreal*

PRO 6: High Performance Fiber Reinforced Cement Composites—HPFRCC 3 (ISBN: 2-912143-06-3; e-ISBN: 2351580222); *Eds. H. W. Reinhardt and A. E. Naaman*

PRO 7: 1st International RILEM Symposium on Self-Compacting Concrete (ISBN: 2-912143-09-8; e-ISBN: 2912143721); *Eds. Å. Skarendahl and Ö. Petersson*

PRO 8: International RILEM Symposium on Timber Engineering (ISBN: 2-912143-10-1; e-ISBN: 2351580230); *Ed. L. Boström*

PRO 9: 2nd International RILEM Symposium on Adhesion between Polymers and Concrete ISAP '99 (ISBN: 2-912143-11-X; e-ISBN: 2351580249); *Eds. Y. Ohama and M. Puterman*

PRO 10: 3rd International RILEM Symposium on Durability of Building and Construction Sealants (ISBN: 2-912143-13-6; e-ISBN: 2351580257); *Ed. A. T. Wolf*

PRO 11: 4th International RILEM Conference on Reflective Cracking in Pavements (ISBN: 2-912143-14-4; e-ISBN: 2351580265); *Eds. A. O. Abd El Halim, D. A. Taylor and El H. H. Mohamed*

PRO 12: International RILEM Workshop on Historic Mortars: Characteristics and Tests (ISBN: 2-912143-15-2; e-ISBN: 2351580273); *Eds. P. Bartos, C. Groot and J. J. Hughes*

PRO 13: 2nd International RILEM Symposium on Hydration and Setting (ISBN: 2-912143-16-0; e-ISBN: 2351580281); *Ed. A. Nonat*

PRO 14: Integrated Life-Cycle Design of Materials and Structures—ILCDES 2000 (ISBN: 951-758-408-3; e-ISBN: 235158029X); (ISSN: 0356-9403); *Ed. S. Sarja*

PRO 15: Fifth RILEM Symposium on Fibre-Reinforced Concretes (FRC)— BEFIB'2000 (ISBN: 2-912143-18-7; e-ISBN: 291214373X); *Eds. P. Rossi and G. Chanvillard*

PRO 16: Life Prediction and Management of Concrete Structures (ISBN: 2-912143-19-5; e-ISBN: 2351580303); *Ed. D. Naus*

PRO 17: Shrinkage of Concrete—Shrinkage 2000 (ISBN: 2-912143-20-9; e-ISBN: 2351580311); *Eds. V. Baroghel-Bouny and P.-C. Aïtcin*

PRO 18: Measurement and Interpretation of the On-Site Corrosion Rate (ISBN: 2-912143-21-7; e-ISBN: 235158032X); *Eds. C. Andrade, C. Alonso, J. Fullea, J. Polimon and J. Rodriguez*

PRO 19: Testing and Modelling the Chloride Ingress into Concrete (ISBN: 2-912143-22-5; e-ISBN: 2351580338); *Eds. C. Andrade and J. Kropp*

PRO 20: 1st International RILEM Workshop on Microbial Impacts on Building Materials (CD 02) (e-ISBN 978-2-35158-013-4); *Ed. M. Ribas Silva*

PRO 21: International RILEM Symposium on Connections between Steel and Concrete (ISBN: 2-912143-25-X; e-ISBN: 2351580346); *Ed. R. Eligehausen*

PRO 22: International RILEM Symposium on Joints in Timber Structures (ISBN: 2-912143-28-4; e-ISBN: 2351580354); *Eds. S. Aicher and H.-W. Reinhardt*

PRO 23: International RILEM Conference on Early Age Cracking in Cementitious Systems (ISBN: 2-912143-29-2; e-ISBN: 2351580362); *Eds. K. Kovler and A. Bentur*

PRO 24: 2nd International RILEM Workshop on Frost Resistance of Concrete (ISBN: 2-912143-30-6; e-ISBN: 2351580370); *Eds. M. J. Setzer, R. Auberg and H.-J. Keck*

PRO 25: International RILEM Workshop on Frost Damage in Concrete (ISBN: 2-912143-31-4; e-ISBN: 2351580389); *Eds. D. J. Janssen, M. J. Setzer and M. B. Snyder*

PRO 26: International RILEM Workshop on On-Site Control and Evaluation of Masonry Structures (ISBN: 2-912143-34-9; e-ISBN: 2351580141); *Eds. L. Binda and R. C. de Vekey*

PRO 27: International RILEM Symposium on Building Joint Sealants (CD03; e-ISBN: 235158015X); *Ed. A. T. Wolf*

PRO 28: 6th International RILEM Symposium on Performance Testing and Evaluation of Bituminous Materials—PTEBM'03 (ISBN: 2-912143-35-7; e-ISBN: 978-2-912143-77-8); *Ed. M. N. Partl*

PRO 29: 2nd International RILEM Workshop on Life Prediction and Ageing Management of Concrete Structures (ISBN: 2-912143-36-5; e-ISBN: 2912143780); *Ed. D. J. Naus*

PRO 30: 4th International RILEM Workshop on High Performance Fiber Reinforced Cement Composites—HPFRCC 4 (ISBN: 2-912143-37-3; e-ISBN: 2912143799); *Eds. A. E. Naaman and H. W. Reinhardt*

PRO 31: International RILEM Workshop on Test and Design Methods for Steel Fibre Reinforced Concrete: Background and Experiences (ISBN: 2-912143-38-1; e-ISBN: 2351580168); *Eds. B. Schnütgen and L. Vandewalle*

PRO 32: International Conference on Advances in Concrete and Structures 2 vol. (ISBN (set): 2-912143-41-1; e-ISBN: 2351580176); *Eds. Ying-shu Yuan, Surendra P. Shah and Heng-lin Lü*

PRO 33: 3rd International Symposium on Self-Compacting Concrete (ISBN: 2-912143-42-X; e-ISBN: 2912143713); *Eds. Ó. Wallevik and I. Nielsson*

PRO 34: International RILEM Conference on Microbial Impact on Building Materials (ISBN: 2-912143-43-8; e-ISBN: 2351580184); *Ed. M. Ribas Silva*

PRO 35: International RILEM TC 186-ISA on Internal Sulfate Attack and Delayed Ettringite Formation (ISBN: 2-912143-44-6; e-ISBN: 2912143802); *Eds. K. Scrivener and J. Skalny*

PRO 36: International RILEM Symposium on Concrete Science and Engineering —A Tribute to Arnon Bentur (ISBN: 2-912143-46-2; e-ISBN: 2912143586); *Eds. K. Kovler, J. Marchand, S. Mindess and J. Weiss*

PRO 37: 5th International RILEM Conference on Cracking in Pavements— Mitigation, Risk Assessment and Prevention (ISBN: 2-912143-47-0; e-ISBN: 2912143764); *Eds. C. Petit, I. Al-Qadi and A. Millien*

PRO 38: 3rd International RILEM Workshop on Testing and Modelling the Chloride Ingress into Concrete (ISBN: 2-912143-48-9; e-ISBN: 2912143578); *Eds. C. Andrade and J. Kropp*

PRO 39: 6th International RILEM Symposium on Fibre-Reinforced Concretes—BEFIB 2004 (ISBN: 2-912143-51-9; e-ISBN: 2912143748); *Eds. M. Di Prisco, R. Felicetti and G. A. Plizzari*

PRO 40: International RILEM Conference on the Use of Recycled Materials in Buildings and Structures (ISBN: 2-912143-52-7; e-ISBN: 2912143756); *Eds. E. Vázquez, Ch. F. Hendriks and G. M. T. Janssen*

PRO 41: RILEM International Symposium on Environment-Conscious Materials and Systems for Sustainable Development (ISBN: 2-912143-55-1; e-ISBN: 2912143640); *Eds. N. Kashino and Y. Ohama*

PRO 42: SCC'2005—China: 1st International Symposium on Design, Performance and Use of Self-Consolidating Concrete (ISBN: 2-912143-61-6; e-ISBN: 2912143624); *Eds. Zhiwu Yu, Caijun Shi, Kamal Henri Khayat and Youjun Xie*

PRO 43: International RILEM Workshop on Bonded Concrete Overlays (e-ISBN: 2-912143-83-7); *Eds. J. L. Granju and J. Silfwerbrand*

PRO 44: 2nd International RILEM Workshop on Microbial Impacts on Building Materials (CD11) (e-ISBN: 2-912143-84-5); *Ed. M. Ribas Silva*

PRO 45: 2nd International Symposium on Nanotechnology in Construction, Bilbao (ISBN: 2-912143-87-X; e-ISBN: 2912143888); *Eds. Peter J. M. Bartos, Yolanda de Miguel and Antonio Porro*

PRO 46: Concrete Life'06—International RILEM-JCI Seminar on Concrete Durability and Service Life Planning: Curing, Crack Control, Performance in Harsh Environments (ISBN: 2-912143-89-6; e-ISBN: 291214390X); *Ed. K. Kovler*

PRO 47: International RILEM Workshop on Performance Based Evaluation and Indicators for Concrete Durability (ISBN: 978-2-912143-95-2; e-ISBN: 9782912143969); *Eds. V. Baroghel-Bouny, C. Andrade, R. Torrent and K. Scrivener*

PRO 48: 1st International RILEM Symposium on Advances in Concrete through Science and Engineering (e-ISBN: 2-912143-92-6); *Eds. J. Weiss, K. Kovler, J. Marchand, and S. Mindess*

PRO 49: International RILEM Workshop on High Performance Fiber Reinforced Cementitious Composites in Structural Applications (ISBN: 2-912143-93-4; e-ISBN: 2912143942); *Eds. G. Fischer and V. C. Li*

PRO 50: 1st International RILEM Symposium on Textile Reinforced Concrete (ISBN: 2-912143-97-7; e-ISBN: 2351580087); *Eds. Josef Hegger, Wolfgang Brameshuber and Norbert Will*

PRO 51: 2nd International Symposium on Advances in Concrete through Science and Engineering (ISBN: 2-35158-003-6; e-ISBN: 2-35158-002-8); *Eds. J. Marchand, B. Bissonnette, R. Gagné, M. Jolin and F. Paradis*

PRO 52: Volume Changes of Hardening Concrete: Testing and Mitigation (ISBN: 2-35158-004-4; e-ISBN: 2-35158-005-2); *Eds. O. M. Jensen, P. Lura and K. Kovler*

PRO 53: High Performance Fiber Reinforced Cement Composites—HPFRCC5 (ISBN: 978-2-35158-046-2; e-ISBN: 978-2-35158-089-9); *Eds. H. W. Reinhardt and A. E. Naaman*

PRO 54: 5th International RILEM Symposium on Self-Compacting Concrete (ISBN: 978-2-35158-047-9; e-ISBN: 978-2-35158-088-2); *Eds. G. De Schutter and V. Boel*

PRO 55: International RILEM Symposium Photocatalysis, Environment and Construction Materials (ISBN: 978-2-35158-056-1; e-ISBN: 978-2-35158-057-8); *Eds. P. Baglioni and L. Cassar*

PRO 56: International RILEM Workshop on Integral Service Life Modelling of Concrete Structures (ISBN 978-2-35158-058-5; e-ISBN: 978-2-35158-090-5); *Eds. R. M. Ferreira, J. Gulikers and C. Andrade*

PRO 57: RILEM Workshop on Performance of cement-based materials in aggressive aqueous environments (e-ISBN: 978-2-35158-059-2); *Ed. N. De Belie*

PRO 58: International RILEM Symposium on Concrete Modelling—CONMOD'08 (ISBN: 978-2-35158-060-8; e-ISBN: 978-2-35158-076-9); *Eds. E. Schlangen and G. De Schutter*

PRO 59: International RILEM Conference on On Site Assessment of Concrete, Masonry and Timber Structures—SACoMaTiS 2008 (ISBN set: 978-2-35158-061-5; e-ISBN: 978-2-35158-075-2); *Eds. L. Binda, M. di Prisco and R. Felicetti*

PRO 60: Seventh RILEM International Symposium on Fibre Reinforced Concrete: Design and Applications—BEFIB 2008 (ISBN: 978-2-35158-064-6; e-ISBN: 978-2-35158-086-8); *Ed. R. Gettu*

PRO 61: 1st International Conference on Microstructure Related Durability of Cementitious Composites 2 vol., (ISBN: 978-2-35158-065-3; e-ISBN: 978-2-35158-084-4); *Eds. W. Sun, K. van Breugel, C. Miao, G. Ye and H. Chen*

PRO 62: NSF/ RILEM Workshop: In-situ Evaluation of Historic Wood and Masonry Structures (e-ISBN: 978-2-35158-068-4); *Eds. B. Kasal, R. Anthony and M. Drdácký*

PRO 63: Concrete in Aggressive Aqueous Environments: Performance, Testing and Modelling, 2 vol., (ISBN: 978-2-35158-071-4; e-ISBN: 978-2-35158-082-0); *Eds. M. G. Alexander and A. Bertron*

PRO 64: Long Term Performance of Cementitious Barriers and Reinforced Concrete in Nuclear Power Plants and Waste Management—NUCPERF 2009 (ISBN: 978-2-35158-072-1; e-ISBN: 978-2-35158-087-5); *Eds. V. L'Hostis, R. Gens and C. Gallé*

PRO 65: Design Performance and Use of Self-consolidating Concrete—SCC'2009 (ISBN: 978-2-35158-073-8; e-ISBN: 978-2-35158-093-6); *Eds. C. Shi, Z. Yu, K. H. Khayat and P. Yan*

PRO 66: 2nd International RILEM Workshop on Concrete Durability and Service Life Planning—ConcreteLife'09 (ISBN: 978-2-35158-074-5; ISBN: 978-2-35158-074-5); *Ed. K. Kovler*

PRO 67: Repairs Mortars for Historic Masonry (e-ISBN: 978-2-35158-083-7); *Ed. C. Groot*

PRO 68: Proceedings of the 3rd International RILEM Symposium on 'Rheology of Cement Suspensions such as Fresh Concrete (ISBN 978-2-35158-091-2; e-ISBN: 978-2-35158-092-9); *Eds. O. H. Wallevik, S. Kubens and S. Oesterheld*

PRO 69: 3rd International PhD Student Workshop on 'Modelling the Durability of Reinforced Concrete (ISBN: 978-2-35158-095-0); *Eds. R. M. Ferreira, J. Gulikers and C. Andrade*

PRO 70: 2nd International Conference on 'Service Life Design for Infrastructure' (ISBN set: 978-2-35158-096-7, e-ISBN: 978-2-35158-097-4); *Eds. K. van Breugel, G. Ye and Y. Yuan*

PRO 71: Advances in Civil Engineering Materials—The 50-year Teaching Anniversary of Prof. Sun Wei' (ISBN: 978-2-35158-098-1; e-ISBN: 978-2-35158-099-8); *Eds. C. Miao, G. Ye and H. Chen*

PRO 72: First International Conference on 'Advances in Chemically-Activated Materials—CAM'2010' (2010), 264 pp., ISBN: 978-2-35158-101-8; e-ISBN: 978-2-35158-115-5; *Eds. Caijun Shi and Xiaodong Shen*

PRO 73: 2nd International Conference on 'Waste Engineering and Management—ICWEM 2010' (2010), 894 pp., ISBN: 978-2-35158-102-5; e-ISBN: 978-2-35158-103-2, *Eds. J. Zh. Xiao, Y. Zhang, M. S. Cheung and R. Chu*

PRO 74: International RILEM Conference on 'Use of Superabsorsorbent Polymers and Other New Addditives in Concrete' (2010) 374 pp., ISBN: 978-2-35158-104-9; e-ISBN: 978-2-35158-105-6; *Eds. O.M. Jensen, M.T. Hasholt, and S. Laustsen*

PRO 75: International Conference on 'Material Science—2nd ICTRC—Textile Reinforced Concrete—Theme 1' (2010) 436 pp., ISBN: 978-2-35158-106-3; e-ISBN: 978-2-35158-107-0; *Ed. W. Brameshuber*

PRO 76: International Conference on 'Material Science—HetMat—Modelling of Heterogeneous Materials—Theme 2' (2010) 255 pp., ISBN: 978-2-35158-108-7; e-ISBN: 978-2-35158-109-4; *Ed. W. Brameshuber*

PRO 77: International Conference on 'Material Science—AdIPoC—Additions Improving Properties of Concrete—Theme 3' (2010) 459 pp., ISBN: 978-2-35158-110-0; e-ISBN: 978-2-35158-111-7; *Ed. W. Brameshuber*

PRO 78: 2nd Historic Mortars Conference and RILEM TC 203-RHM Final Workshop—HMC2010 (2010) 1416 pp., e-ISBN: 978-2-35158-112-4; *Eds. J. Válek, C. Groot and J. J. Hughes*

PRO 79: International RILEM Conference on Advances in Construction Materials Through Science and Engineering (2011) 213 pp., ISBN: 978-2-35158-116-2, e-ISBN: 978-2-35158-117-9; *Eds. Christopher Leung and K.T. Wan*

PRO 80: 2nd International RILEM Conference on Concrete Spalling due to Fire Exposure (2011) 453 pp., ISBN: 978-2-35158-118-6; e-ISBN: 978-2-35158-119-3; *Eds. E.A.B. Koenders and F. Dehn*

PRO 81: 2nd International RILEM Conference on Strain Hardening Cementitious Composites (SHCC2-Rio) (2011) 451 pp., ISBN: 978-2-35158-120-9; e-ISBN: 978-2-35158-121-6; *Eds. R.D. Toledo Filho, F.A. Silva, E.A.B. Koenders and E.M.R. Fairbairn*

PRO 82: 2nd International RILEM Conference on Progress of Recycling in the Built Environment (2011) 507 pp., e-ISBN: 978-2-35158-122-3; *Eds. V.M. John, E. Vazquez, S.C. Angulo and C. Ulsen*

PRO 83: 2nd International Conference on Microstructural-related Durability of Cementitious Composites (2012) 250 pp., ISBN: 978-2-35158-129-2; e-ISBN: 978-2-35158-123-0; *Eds. G. Ye, K. van Breugel, W. Sun and C. Miao*

PRO 84: CONSEC13—Seventh International Conference on Concrete under Severe Conditions—Environment and Loading (2013) 1930 pp., ISBN: 978-2-35158-124-7; e-ISBN: 978-2- 35158-134-6; *Eds. Z.J. Li, W. Sun, C.W. Miao, K. Sakai, O.E. Gjorv and N. Banthia*

PRO 85: RILEM-JCI International Workshop on Crack Control of Mass Concrete and Related issues concerning Early-Age of Concrete Structures—ConCrack 3— Control of Cracking in Concrete Structures 3 (2012) 237 pp., ISBN: 978-2-35158-125-4; e-ISBN: 978-2-35158-126-1; *Eds. F. Toutlemonde and J.-M. Torrenti*

PRO 86: International Symposium on Life Cycle Assessment and Construction (2012) 414 pp., ISBN: 978-2-35158-127-8, e-ISBN: 978-2-35158-128-5; *Eds. A. Ventura and C. de la Roche*

PRO 87: UHPFRC 2013—RILEM-fib-AFGC International Symposium on Ultra-High Performance Fibre-Reinforced Concrete (2013), ISBN: 978-2-35158-130-8, e-ISBN: 978-2-35158-131-5; *Eds. F. Toutlemonde*

PRO 88: 8th RILEM International Symposium on Fibre Reinforced Concrete (2012) 344 pp., ISBN: 978-2-35158-132-2; e-ISBN: 978-2-35158-133-9; *Eds. Joaquim A.O. Barros*

PRO 89: RILEM International workshop on performance-based specification and control of concrete durability (2014) 678 pp., ISBN: 978-2-35158-135-3; e-ISBN: 978-2-35158-136-0; *Eds. D. Bjegović, H. Beushausen and M. Serdar*

PRO 90: 7th RILEM International Conference on Self-Compacting Concrete and of the 1st RILEM International Conference on Rheology and Processing of Construction Materials (2013) 396 pp., ISBN: 978-2-35158-137-7; e-ISBN: 978-2-35158-138-4; *Eds. Nicolas Roussel and Hela Bessaies-Bey*

PRO 91: CONMOD 2014—RILEM International Symposium on Concrete Modelling (2014), ISBN: 978-2-35158-139-1; e-ISBN: 978-2-35158-140-7; *Eds. Kefei Li, Peiyu Yan and Rongwei Yang*

PRO 92: CAM 2014—2nd International Conference on advances in chemically-activated materials (2014) 392 pp., ISBN: 978-2-35158-141-4; e-ISBN: 978-2-35158-142-1; *Eds. Caijun Shi and Xiadong Shen*

PRO 93: SCC 2014—3rd International Symposium on Design, Performance and Use of Self-Consolidating Concrete (2014) 438 pp., ISBN: 978-2-35158-143-8; e-ISBN: 978-2-35158-144-5; *Eds. Caijun Shi, Zhihua Ou and Kamal H. Khayat*

PRO 94 (online version): HPFRCC-7—7th RILEM conference on High performance fiber reinforced cement composites (2015), e-ISBN: 978-2-35158-146-9; *Eds. H.W. Reinhardt, G.J. Parra-Montesinos and H. Garrecht*

PRO 95: International RILEM Conference on Application of superabsorbent polymers and other new admixtures in concrete construction (2014), ISBN: 978-2-35158-147-6; e-ISBN: 978-2-35158-148-3; *Eds. Viktor Mechtcherine and Christof Schroefl*

PRO 96 (online version): XIII DBMC: XIII International Conference on Durability of Building Materials and Components (2015), e-ISBN: 978-2-35158-149-0; *Eds. M. Quattrone and V.M. John*

PRO 97: SHCC3—3rd International RILEM Conference on Strain Hardening Cementitious Composites (2014), ISBN: 978-2-35158-150-6; e-ISBN: 978-2-35158-151-3; *Eds. E. Schlangen, M.G. Sierra Beltran, M. Lukovic and G. Ye*

PRO 98: FERRO-11—11th International Symposium on Ferrocement and 3rd ICTRC—International Conference on Textile Reinforced Concrete (2015), ISBN: 978-2-35158-152-0; e-ISBN: 978-2-35158-153-7; *Ed. W. Brameshuber*

PRO 99 (online version): ICBBM 2015—1st International Conference on Bio-Based Building Materials (2015), e-ISBN: 978-2-35158-154-4; *Eds. S. Amziane and M. Sonebi*

PRO 100: SCC16—RILEM Self-Consolidating Concrete Conference (2016), ISBN: 978-2-35158-156-8; e-ISBN: 978-2-35158-157-5; *Ed. Kamal H. Kayat*

PRO 101 (online version): III Progress of Recycling in the Built Environment (2015), e-ISBN: 978-2-35158-158-2; *Eds I. Martins, C. Ulsen and S. C. Angulo*

PRO 102 (online version): RILEM Conference on Microorganisms-Cementitious Materials Interactions (2016), e-ISBN: 978-2-35158-160-5; *Eds. Alexandra Bertron, Henk Jonkers and Virginie Wiktor*

PRO 103 (online version): ACESC'16—Advances in Civil Engineering and Sustainable Construction (2016), e-ISBN: 978-2-35158-161-2; *Eds. T.Ch. Madhavi, G. Prabhakar, Santhosh Ram and P.M. Rameshwaran*

PRO 104 (online version): SSCS'2015—Numerical Modeling—Strategies for Sustainable Concrete Structures (2015), e-ISBN: 978-2-35158-162-9

PRO 105: 1st International Conference on UHPC Materials and Structures (2016), ISBN: 978-2-35158-164-3; e-ISBN: 978-2-35158-165-0

PRO 106: AFGC-ACI-fib-RILEM International Conference on Ultra-High-Performance Fibre-Reinforced Concrete—UHPFRC 2017 (2017), ISBN: 978-2-35158-166-7; e-ISBN: 978-2-35158-167-4; *Eds. François Toutlemonde and Jacques Resplendino*

PRO 107 (online version): XIV DBMC—14th International Conference on Durability of Building Materials and Components (2017), e-ISBN: 978-2-35158-159-9; *Eds. Geert De Schutter, Nele De Belie, Arnold Janssens and Nathan Van Den Bossche*

PRO 108: MSSCE 2016—Innovation of Teaching in Materials and Structures (2016), ISBN: 978-2-35158-178-0; e-ISBN: 978-2-35158-179-7; *Ed. Per Goltermann*

PRO 109 (2 volumes): MSSCE 2016—Service Life of Cement-Based Materials and Structures (2016), ISBN Vol. 1: 978-2-35158-170-4; Vol. 2: 978-2-35158-171-4; Set Vol. 1&2: 978-2-35158-172-8; e-ISBN : 978-2-35158-173-5; *Eds. Miguel Azenha, Ivan Gabrijel, Dirk Schlicke, Terje Kanstad and Ole Mejlhede Jensen*

PRO 110: MSSCE 2016—Historical Masonry (2016), ISBN: 978-2-35158-178-0; e-ISBN: 978-2-35158-179-7; *Eds. Inge Rörig-Dalgaard and Ioannis Ioannou*

PRO 111: MSSCE 2016—Electrochemistry in Civil Engineering (2016); ISBN: 978-2-35158-176-6; e-ISBN: 978-2-35158-177-3; *Ed. Lisbeth M. Ottosen*

PRO 112: MSSCE 2016—Moisture in Materials and Structures (2016), ISBN: 978-2-35158-178-0; e-ISBN: 978-2-35158-179-7; *Eds. Kurt Kielsgaard Hansen, Carsten Rode and Lars-Olof Nilsson*

PRO 113: MSSCE 2016—Concrete with Supplementary Cementitious Materials (2016), ISBN: 978-2-35158-178-0; e-ISBN: 978-2-35158-179-7; *Eds. Ole Mejlhede Jensen, Konstantin Kovler and Nele De Belie*

PRO 114: MSSCE 2016—Frost Action in Concrete (2016), ISBN: 978-2-35158-182-7; e-ISBN: 978-2-35158-183-4; *Eds. Marianne Tange Hasholt, Katja Fridh and R. Doug Hooton*

PRO 115: MSSCE 2016—Fresh Concrete (2016), ISBN: 978-2-35158-184-1; e-ISBN: 978-2-35158-185-8; *Eds. Lars N. Thrane, Claus Pade, Oldrich Svec and Nicolas Roussel*

PRO 116: BEFIB 2016—9th RILEM International Symposium on Fiber Reinforced Concrete (2016), ISBN: 978-2-35158-187-2; e-ISBN: 978-2-35158-186-5; *Eds. N. Banthia, M. di Prisco and S. Soleimani-Dashtaki*

PRO 117: 3rd International RILEM Conference on Microstructure Related Durability of Cementitious Composites (2016), ISBN: 978-2-35158-188-9; e-ISBN: 978-2-35158-189-6; *Eds. Changwen Miao, Wei Sun, Jiaping Liu, Huisu Chen, Guang Ye and Klaas van Breugel*

PRO 118 (4 volumes): International Conference on Advances in Construction Materials and Systems (2017), ISBN Set: 978-2-35158-190-2; Vol. 1: 978-2-35158-193-3; Vol. 2: 978-2-35158-194-0; Vol. 3: ISBN:978-2-35158-195-7; Vol. 4: ISBN:978-2-35158-196-4; e-ISBN: 978-2-35158-191-9; *Ed. Manu Santhanam*

PRO 119 (online version): ICBBM 2017—Second International RILEM Conference on Bio-based Building Materials, (2017), e-ISBN: 978-2-35158-192-6; *Ed. Sofiane Amziane*

PRO 120 (2 volumes): EAC-02—2nd International RILEM/COST Conference on Early Age Cracking and Serviceability in Cement-based Materials and Structures, (2017), Vol. 1: 978-2-35158-199-5, Vol. 2: 978-2-35158-200-8, Set: 978-2-35158-197-1, e-ISBN: 978-2-35158-198-8; *Eds. Stéphanie Staquet and Dimitrios Aggelis*

PRO 121 (2 volumes): SynerCrete18: Interdisciplinary Approaches for Cementbased Materials and Structural Concrete: Synergizing Expertise and Bridging Scales of Space and Time, (2018), Set: 978-2-35158-202-2, Vol.1: 978-2-35158-211-4, Vol.2: 978-2-35158-212-1, e-ISBN: 978-2-35158-203-9; *Eds. Miguel Azenha, Dirk Schlicke, Farid Benboudjema, Agnieszka Knoppik*

PRO 122: SCC'2018 China—Fourth International Symposium on Design, Performance and Use of Self-Consolidating Concrete, (2018), ISBN: 978-2-35158-204-6, e-ISBN: 978-2-35158-205-3; *Eds. C. Shi, Z. Zhang, K. H. Khayat*

PRO 123: Final Conference of RILEM TC 253-MCI: Microorganisms-Cementitious Materials Interactions (2018), Set: 978-2-35158-207-7, Vol.1: 978-2-35158-209-1, Vol.2: 978-2-35158-210-7, e-ISBN: 978-2-35158-206-0; *Ed. Alexandra Bertron*

PRO 124 (online version): Fourth International Conference Progress of Recycling in the Built Environment (2018), e-ISBN: 978-2-35158-208-4; *Eds. Isabel M. Martins, Carina Ulsen, Yury Villagran*

PRO 125 (online version): SLD4—4th International Conference on Service Life Design for Infrastructures (2018), e-ISBN: 978-2-35158-213-8; *Eds. Guang Ye, Yong Yuan, Claudia Romero Rodriguez, Hongzhi Zhang, Branko Savija*

PRO 126: Workshop on Concrete Modelling and Material Behaviour in honor of Professor Klaas van Breugel (2018), ISBN: 978-2-35158-214-5, e-ISBN: 978-2-35158-215-2; *Ed. Guang Ye*

PRO 127 (online version): CONMOD2018—Symposium on Concrete Modelling (2018), e-ISBN: 978-2-35158-216-9; *Eds. Erik Schlangen, Geert de Schutter, Branko Savija, Hongzhi Zhang, Claudia Romero Rodriguez*

PRO 128: SMSS2019—International Conference on Sustainable Materials, Systems and Structures (2019), ISBN: 978-2-35158-217-6, e-ISBN: 978-2-35158-218-3

PRO 129: 2nd International Conference on UHPC Materials and Structures (UHPC2018-China), ISBN: 978-2-35158-219-0, e-ISBN: 978-2-35158-220-6

PRO 130: 5th Historic Mortars Conference (2019), ISBN: 978-2-35158-221-3, e-ISBN: 978-2-35158-222-0; *Eds. José Ignacio Álvarez, José María Fernández, Íñigo Navarro, Adrián Durán, Rafael Sirera*

PRO 131 (online version): 3rd International Conference on Bio-Based Building Materials (ICBBM2019), e-ISBN: 978-2-35158-229-9; *Eds. Mohammed Sonebi, Sofiane Amziane, Jonathan Page*

PRO 132: IRWRMC'18—International RILEM Workshop on Rheological Measurements of Cement-based Materials (2018), ISBN: 978-2-35158-230-5, e-ISBN: 978-2-35158-231-2; *Eds. Chafika Djelal, Yannick Vanhove*

PRO 133 (online version): CO2STO2019—International Workshop CO2 Storage in Concrete (2019), e-ISBN: 978-2-35158-232-9; *Eds. Assia Djerbi, Othman Omikrine-Metalssi, Teddy Fen-Chong*

RILEM Reports (REP)

Report 19: Considerations for Use in Managing the Aging of Nuclear Power Plant Concrete Structures (ISBN: 2-912143-07-1); *Ed. D. J. Naus*

Report 20: Engineering and Transport Properties of the Interfacial Transition Zone in Cementitious Composites (ISBN: 2-912143-08-X); *Eds. M. G. Alexander, G. Arliguie, G. Ballivy, A. Bentur and J. Marchand*

Report 21: Durability of Building Sealants (ISBN: 2-912143-12-8); *Ed. A. T. Wolf*

Report 22: Sustainable Raw Materials—Construction and Demolition Waste (ISBN: 2-912143-17-9); *Eds. C. F. Hendriks and H. S. Pietersen*

Report 23: Self-Compacting Concrete state-of-the-art report (ISBN: 2-912143-23-3); *Eds. Å. Skarendahl and Ö. Petersson*

Report 24: Workability and Rheology of Fresh Concrete: Compendium of Tests (ISBN: 2-912143-32-2); *Eds. P. J. M. Bartos, M. Sonebi and A. K. Tamimi*

Report 25: Early Age Cracking in Cementitious Systems (ISBN: 2-912143-33-0); *Ed. A. Bentur*

Report 26: Towards Sustainable Roofing (Joint Committee CIB/RILEM) (CD 07) (e-ISBN 978-2-912143-65-5); *Eds. Thomas W. Hutchinson and Keith Roberts*

Report 27: Condition Assessment of Roofs (Joint Committee CIB/RILEM) (CD 08) (e-ISBN 978-2-912143-66-2); *Ed. CIB W 83/RILEM TC166-RMS*

Report 28: Final report of RILEM TC 167-COM 'Characterisation of Old Mortars with Respect to Their Repair (ISBN: 978-2-912143-56-3); *Eds. C. Groot, G. Ashall and J. Hughes*

Report 29: Pavement Performance Prediction and Evaluation (PPPE): Interlaboratory Tests (e-ISBN: 2-912143-68-3); *Eds. M. Partl and H. Piber*

Report 30: Final Report of RILEM TC 198-URM 'Use of Recycled Materials' (ISBN: 2-912143-82-9; e-ISBN: 2-912143-69-1); *Eds. Ch. F. Hendriks, G. M. T. Janssen and E. Vázquez*

Report 31: Final Report of RILEM TC 185-ATC 'Advanced testing of cement-based materials during setting and hardening' (ISBN: 2-912143-81-0; e-ISBN: 2-912143-70-5); *Eds. H. W. Reinhardt and C. U. Grosse*

Report 32: Probabilistic Assessment of Existing Structures. A JCSS publication (ISBN 2-912143-24-1); *Ed. D. Diamantidis*

Report 33: State-of-the-Art Report of RILEM Technical Committee TC 184-IFE 'Industrial Floors' (ISBN 2-35158-006-0); *Ed. P. Seidler*

Report 34: Report of RILEM Technical Committee TC 147-FMB 'Fracture mechanics applications to anchorage and bond' Tension of Reinforced Concrete Prisms—Round Robin Analysis and Tests on Bond (e-ISBN 2-912143-91-8); *Eds. L. Elfgren and K. Noghabai*

Report 35: Final Report of RILEM Technical Committee TC 188-CSC 'Casting of Self Compacting Concrete' (ISBN 2-35158-001-X; e-ISBN: 2-912143-98-5); *Eds. Å. Skarendahl and P. Billberg*

Report 36: State-of-the-Art Report of RILEM Technical Committee TC 201-TRC 'Textile Reinforced Concrete' (ISBN 2-912143-99-3); *Ed. W. Brameshuber*

Report 37: State-of-the-Art Report of RILEM Technical Committee TC 192-ECM 'Environment-conscious construction materials and systems' (ISBN: 978-2-35158-053-0); *Eds. N. Kashino, D. Van Gemert and K. Imamoto*

Report 38: State-of-the-Art Report of RILEM Technical Committee TC 205-DSC 'Durability of Self-Compacting Concrete' (ISBN: 978-2-35158-048-6); *Eds. G. De Schutter and K. Audenaert*

Report 39: Final Report of RILEM Technical Committee TC 187-SOC 'Experimental determination of the stress-crack opening curve for concrete in tension' (ISBN 978-2-35158-049-3); *Ed. J. Planas*

Report 40: State-of-the-Art Report of RILEM Technical Committee TC 189-NEC 'Non-Destructive Evaluation of the Penetrability and Thickness of the Concrete Cover' (ISBN 978-2-35158-054-7); *Eds. R. Torrent and L. Fernández Luco*

Report 41: State-of-the-Art Report of RILEM Technical Committee TC 196-ICC 'Internal Curing of Concrete' (ISBN 978-2-35158-009-7); *Eds. K. Kovler und O. M. Jensen*

Report 42: 'Acoustic Emission and Related Non-destructive Evaluation Techniques for Crack Detection and Damage Evaluation in Concrete'—Final Report of RILEM Technical Committee 212-ACD (e-ISBN: 978-2-35158-100-1); *Ed. M. Ohtsu*

Report 45: Repair Mortars for Historic Masonry—State-of-the-Art Report of RILEM Technical Committee TC 203-RHM (e-ISBN: 978-2-35158-163-6); *Eds. Paul Maurenbrecher and Caspar Groot*

Report 46: Surface delamination of concrete industrial ffioors and other durability related aspects guide—Report of RILEM Technical Committee TC 268-SIF (e-ISBN: 978-2-35158-201-5); *Ed. Valerie Pollet*

Contents

9   Statements and Conclusions . . . .
   Aitor Llano-Torre, Pedro Serna, Sergio H. P. Cavalaro, Nicola Buratti,
   E. Stefan Bernard, William P. Boshoff, Raúl L. Zerbino, Hans Pauwels,
   Wolfgang Kusterle, Emilio Garcia-Taengua, Rutger Vrijdaghs,
   Clementina del Prete, Karyne F. dos Santos, Benoît Parmentier,
   and Claudio Mazzotti

Symbols

FRC	Fibre-reinforced concrete
SyFRC	Synthetic fibre-reinforced concrete
SFRC	Steel fibre-reinforced concrete
FRS	Fibre-reinforced shotcrete
CMOD	Crack mouth opening displacement
COD	Crack opening displacement
δ	Deflection
CTOD	Crack tip opening displacement
$CMOD_{pn}$	Target nominal CMOD in the pre-cracking stage
$CMOD_p$	Maximum CMOD reached in the pre-cracking stage
$CMOD_{pri}$	Elastic CMOD recovery after unloading in the pre-cracking stage
$CMOD_{pr}$	Residual CMOD 10 min after unloading in the pre-cracking stage
$CMOD_{ci}$	Instantaneous CMOD immediately after reaching the reference load (Point E in Fig. 4.1)
$CMOD_{ci}^{10'}$	Short-term CMOD 10 min after reaching the reference load (Point E+10' in Fig. 4.1)
$CMOD_{ci}^{30'}$	Short-term CMOD 30 min after reaching the reference load (Point E+30' in Fig. 4.1)
$CMOD_{cd}^{j}$	Delayed CMOD after j days in the creep test
$CMOD_{ct}^{j}$	Total CMOD after j days in the creep test (sum of instantaneous and delayed CMOD)
$CMOD_{cri}$	Elastic CMOD recovery after unloading the creep test
$CMOD_{crd}$	Delayed CMOD recovery 30 days after unloading the creep test
$CMOD_o$	Absolute CMOD assessed from the origin of deformations (Point O in Fig. 4.1)
$CMOD_{oi}$	Absolute instantaneous CMOD assessed from the origin of deformations (Point O in Fig. 4.1)
$CMOD_o^{j}$	Absolute CMOD after j days assessed from the origin of deformations (Point O in Fig. 4.1)

$\varphi_{w,c}{}^{j}$	Crack opening creep coefficient referring to creep stage at j days
$\varphi_{w,o}{}^{j}$	Crack opening creep coefficient referring to origin at j days
$COR^{j\text{-}k}$	Crack opening rate between j and k days
f_L	Residual flexural tensile strength at the limit of proportionality (LOP)
$f_{R,1}$	Residual flexural tensile strength corresponding to $CMOD_1 = 0.5$ mm
$f_{R,2}$	Residual flexural tensile strength corresponding to $CMOD_2 = 1.5$ mm
$f_{R,3}$	Residual flexural tensile strength corresponding to $CMOD_3 = 2.5$ mm
$f_{R,4}$	Residual flexural tensile strength corresponding to $CMOD_4 = 3.5$ mm
$f_{PostCreep,R,2}$	Residual flexural tensile strength corresponding to origin at $CMOD_2 = 1.5$ mm
$f_{PostCreep,R,3}$	Residual flexural tensile strength corresponding to origin at $CMOD_3 = 2.5$ mm
$f_{PostCreep,R,4}$	Residual flexural tensile strength corresponding to origin at $CMOD_4 = 3.5$ mm
$f_{R,p}$	Residual flexural tensile strength at $CMOD_p$
$F_{R,p}$	Load corresponding to $CMOD_p$
$f_{R,c}$	Stress applied during the creep stage
$F_{R,c}$	Load applied to the specimen during the creep stage
F_L	Load at the limit of proportionality (LOP)
F_1	Load corresponding to $CMOD_1 = 0.5$ mm
$F_{PostCreep,2}$	Load corresponding to $CMOD_2 = 1.5$ mm
$F_{PostCreep,3}$	Load corresponding to $CMOD_3 = 2.5$ mm
$F_{PostCreep,4}$	Load corresponding to $CMOD_4 = 3.5$ mm
t_{ci}	Time duration of the loading process in the creep test
t_{cri}	Time duration of the unloading process in the creep test
t_{crd}	Time duration after unloading of the creep test in which recovery was registered
I_n	Nominal creep index or stress level as percentage of $f_{R,p}$
I_c	Applied creep index or stress level as percentage of $f_{R,p}$ ($I_c = f_{R,c}/f_{R,p}$)
n_f	Average number of fibres per cm^2 of fracture surface area ($fibres/cm^2$)
L	Span between supports in flexural test
l_a	Distance between support and nearest loading point in the flexural test
l_b	Span between loading points in the flexural test

CV   Coefficient of variation
LOP   Limit of proportionality

Note Displacement notations are defined only for crack mouth opening displacement (CMOD). Analogous notations also apply to the crack opening displacement (COD) or deflection (δ). The analogous symbols and definitions are obtained substituting the CMOD by COD or δ.

Chapter 1
Introduction and Background

Aitor Llano-Torre, Pedro Serna, Sergio H. P. Cavalaro, Nicola Buratti,
E. Stefan Bernard, William P. Boshoff, Raúl L. Zerbino, Hans Pauwels,
Wolfgang Kusterle, Emilio Garcia-Taengua, Rutger Vrijdaghs,
Clementina del Prete, Karyne F. dos Santos, Benoît Parmentier,
and Claudio Mazzotti

Abstract The absence of a standardised methodology to evaluate creep in the
cracked state of fibre-reinforced concrete (FRC) hindered general comparisons and
conclusions that could lead to significant advances in this topic. A coordinated effort
was required to improve the knowledge on long-term behaviour of cracked FRC

A. Llano-Torre (✉) · P. Serna
Institute of Concrete Science and Technology ICITECH, Universitat Politècnica de València
(UPV), Valencia, Spain
e-mail: aillator@upv.es

P. Serna
e-mail: pserna@cst.upv.es

S. H. P. Cavalaro
School of Architecture, Building and Civil Engineering, Loughborough University,
Loughborough, UK
e-mail: s.cavalaro@lboro.ac.uk

N. Buratti · C. del Prete · C. Mazzotti
Department of Civil, Chemical, Environmental and Materials Engineering DICAM, University of
Bologna, Bologna, Italy
e-mail: nicola.buratti@unibo.it

C. del Prete
e-mail: clementina.delprete2@unibo.it

C. Mazzotti
e-mail: claudio.mazzotti@unibo.it

E. Stefan Bernard
TSE Technologies in Structural Engineering Pty Ltd, Sydney, Australia
e-mail: s.bernard@tse.net.au

W. P. Boshoff
Faculty of Engineering, Built Environment and Information Technology, University of Pretoria,
Pretoria, South Africa
e-mail: billy.boshoff@up.ac.za

R. L. Zerbino
LEMIT-CIC and Faculty of Engineering UNLP, La Plata, Argentina
e-mail: zerbino@ing.unlp.edu.ar

© RILEM 2021
A. Llano-Torre and P. Serna (eds.), *Round-Robin Test on Creep Behaviour
in Cracked Sections of FRC: Experimental Program, Results and Database Analysis*,
RILEM State-of-the-Art Reports 34, https://doi.org/10.1007/978-3-030-72736-9_1

1

sections and assess the different testing methodologies available. The RILEM Technical Committee 261-CCF launched in 2015 the round-robin test (RRT) on creep behaviour in cracked sections of FRC program. This chapter includes the introduction and background to understand the realisation of this RRT. The main objectives as well as the limitations of the scope of the RRT are explained. The structure of the book is presented to provide a first overview of the book content. Finally, information about institutions involved as well as geographic distribution of RRT participants and number of specimens tested in the RRT is provided.

The introduction of fibre-reinforced concrete (FRC) in the world of concrete is a fact for many years, and the use of FRC in the construction of pavements or tunnel linings is widely extended. On the contrary, this material has been very late included in structural standards and designing codes. The Model Code [1] first incorporated the contribution of fibres in mechanical behaviour in 2010. The American Concrete Institute also introduced some FRC contributions in ACI 318-14 [2]. Afterwards, other national codes have also incorporated this option as concrete reinforcement and a similar update is expected in the next version of the Eurocode EC2.

The knowledge of the creep response is essential to ensure long-term performance and safety. The interest of the scientific community on the long-term behaviour of the FRC in structural applications has rapidly increased in recent years. The long-term behaviour of FRC under sustained load in cracked state is still a controversial topic in both scientific and technical international forums and therefore, it is one of the main issues that hinder a prompt introduction of FRC in structural applications. Although several codes include the recommendations for the design of FRC structures, very

H. Pauwels
NV BEKAERT SA, Zwevegem, Belgium
e-mail: hans.pauwels@bekaert.com

W. Kusterle
OTH Regensburg, Regensburg University of Applied Sciences, Regensburg, Germany
e-mail: wolfgang@kusterle.net

E. Garcia-Taengua
School of Civil Engineering, University of Leeds, Leeds, UK
e-mail: E.Garcia-Taengua@leeds.ac.uk

R. Vrijdaghs
Building Materials and Building Technology Section, KU Leuven, Louvain, Belgium
e-mail: rutger.vrijdaghs@kuleuven.be

K. F. dos Santos
Department of Civil and Environmental Engineering, University of Brasília, Brasília, Brazil
e-mail: karyne.ferreira@aluno.unb.br

B. Parmentier
BBRI Belgium Building Research Institute, Limelette, Belgium
e-mail: benoit.parmentier@bbri.be

low number of studies lead on how to consider the effect of creep on the verifications of service limit state (SLS) [3] and ultimate limit state (ULS) and research on the creep behaviour of FRC is limited to studies with no clear consensus on how to characterise the phenomenon experimentally and the influence of several variables. Studies focused on aspects like the influence of the type of fibre, the fibre-concrete bond or environmental conditions that are difficult to compare if each research group performs a different testing methodology, or the use of different parameters to analyse the phenomenon [4].

As there is not a standardized testing methodology for creep in FRC, several test methods have been developed and proposed with differences in terms of type and size of specimen, loading configuration, test parameters and type of equipment used. The proposals may be grouped in the following categories depending on the internal stress distribution generated: the flexural creep of prisms, flexural creep of panels and direct tension.

The flexural creep test of prismatic specimens usually follows either a 3-point [5] or a 4-point [6, 7] bending setup (3PBT and 4PBT, respectively). Some of these laboratories use a single specimen configuration [8] with one specimen per creep frame to guarantee high accuracy in the load applied, whereas others stack two or three specimens [9] in a column to increase the number of results per frame. Some studies assess the behaviour of unnotched specimens to evaluate the influence of multiple cracking and redistribution, while others characterise notched specimens to induce localisation of a main crack. Flexural creep tests in panels are usually performed in square [10] or round panels [11], following different reference standards. In the case of direct tension, creep tests have been performed with cast specimens (prismatic) [12] or cylindrical cores [13]. Moreover, some laboratories conduct special creep test in splitting properties [14] or test with a sustained load on FRC structural elements [15, 16].

These testing methods assess the contribution of creep under different cross-sectional stress profiles (e.g. compression, un-cracked section in tension, cracked section in tension) that aim to reproduce the conditions found in particular applications, such as tunnels, pavements or structural elements like beams, plates, piers... The broad diversity in testing conditions compromises the definition of a single experimental procedure that could satisfy the needs of all applications.

In the same frame, nowadays there is no numerical approach of such sort for the consideration of creep of FRC in the design of real-scale structures, and there is no more a well-defined creep parameter that would serve as input for the simulation of different types of structures. Although the creep test parameters are usually reported in the literature, few studies have been conducted on the comparison of the results of different methodologies or to evaluate their repeatability and reproducibility. The limited database of results obtained under similar conditions is another barrier to the inclusion of creep in codes, and this is difficult to tackle as any creep test is too much time consuming.

To deal with this issue, the RILEM Technical Committee 261-CCF was created in 2014 to, among other objectives, coordinate research efforts compiling the results of different studies on mechanisms of creep behaviour of FRC in cracked sections. This effort should conclude on agreed methodology recommendations for creep test in cracked state to determine the parameters characterizing long-term behaviour. These recommendations could help to assess the feasibility of those test methods clarifying the criteria to analyse the test results. This was the first required step to be able to analyse in further steps the variables influencing creep (composition and properties of concrete; dosage and fibre type, external conditions of work and stress level).

Performing a round-robin test (RRT) became a useful tool when a common testing method is agreed and the objective is to verify the repeatability and reproducibility of the test, comparing the results of a series of laboratories. In the case of FRC creep behaviour, an initial agreement of a unified methodology was far to be guaranteed due to such numbers of different methodologies for creep test of FRC in cracked state. Therefore, the RILEM TC 261-CCF launched an international RRT program [17] in order to improve the knowledge on long-term behaviour of cracked FRC sections and check all the different testing methodologies, assuming the big variability of testing criteria among the scientific community. A successful call for participants in the RRT program was done in 2014 and most groups working in the topic around the world (19 laboratories from the five continents) decided to join the RRT.

The aim of this RRT was to lay the groundwork for a better understanding the current research position and try to establish the basis for next research, possible standardization, and support the way to compare different concretes or fibres effects on creep. It was created a first work program to be cognizant of the situation of the different research groups that were currently studying this phenomenon more in depth than with the simple reading of scientific publications. Only when the same problem is proposed to different groups, and their response is evaluated, the test capacities and the scope of the procedural differences can be compared.

By proposing the use of their own test methods to all the participants, it was possible to appreciate the test viability and laboratory capacity for each methodology and the purpose for which each method was devised.

1.1   Objectives

The objective of the RRT program was to understand the influence of different testing methods and conditions on the response of FRC specimens produced in a single site and shipped to laboratories around the world. The goal was to identify the variations in results and analysis induced by the different testing methods. The RRT should help identify the main parameters derived from the results of each test method and contribute to reaching a consensus on how creep should be evaluated.

The RRT also provides an extensive database [18] of results from tests performed according to similar procedures and conditions agreed beforehand by the participants. This database will serve as a reference for future research and contribute to the proposal and validation of models to consider creep in the design of FRC structures. The analysis of RRT results should support the future proposal of standard test procedures and conditions, defining common criteria for future experimental programs for each test method.

The main significant objectives of this RRT program were defined as follows:

- Compare the responses of different test methods for identical concrete. This will also provide an indirect view on the repercussion of creep on different types of structures since each test method try to replicate the conditions found in specific real-scale structures.
- Detect differences in the methodology of similar tests and verify whether this may lead to different results and/or interpretations.
- Analyse the repeatability and reproducibility of test methods when performed in different laboratories.
- Evaluate the parameters derived from each test method, reaching a consensus on how creep should be evaluated.
- Generate an extensive database of results from tests performed according with procedures and conditions agreed beforehand by the participants. The analysis of the results obtained should serve to the proposal of standard test conditions and procedures, defining common criteria for future experimental programs with each test method.

1.2 Limitation of the RRT

It is very important also to clearly state the RRT limitations, which were based on discussions held in RILEM TC 261-CCF meetings. The RRT does not aim:

- To evaluate if a type of fibre is better than another in terms of structural performance.
- To propose a single and unified test method that provides results representative of all applications (note that the analysis of RRT results could contribute to future proposals of standard test procedures).
- To generate numerical or conceptual models for the creep behaviour of FRC cracked sections (albeit important, this issue requires further studies from research groups working on the topic).
- To quantify the creep strain expected in specific applications.
- To simulate extreme conditions in terms of loading or cracking (the evaluation of tertiary creep is outside the scope of the RRT and requires a more specific experimental program).

The RRT results and analysis are also limited to the testing conditions described here. They should not be generalised or taken as a reference to predict the creep behaviour of the material out of the tested parameters.

1.3   Structure of the Document

Each RRT participant laboratory proposed the methodology they were ready to perform and capability of specimens to be tested. The different methodologies were grouped in three different sections: *flexural test*, *direct tension test* and *panel test* (for both square and round panels). Initially a total program for testing 142 specimens in creep included 98 flexural creep test, 16 direct tension test, 24 Square panels and 4 round panels with a distribution for laboratory as described in Sect. 1.4.

As the aim of the program was oriented to assess the methodologies more than characterize a series of fibre or FRC, it was decided to use only one concrete matrix. The mix design was defined thinking in the capability to accept a significant dosage of fibres. It was also agreed to analyse the response of the matrix with two different fibres materials (steel and synthetic fibres) in order to verify that the proposed methodologies can detect differences between both with regard to their performance in concrete. The type of fibres used were proposed by the participant fibre manufacturers and the fibre dosages were proposed looking for a structural performance with similar residual strength $f_{R,3}$ for both FRC concretes. The concrete for all creep test and additional specimens related to this program was produced in a precast company facility located close to Valencia. The defined mix design details as well as the details of the production and delivery of the specimens are presented in Chap. 2. Additional information about the index of the produced specimens and their destination can be found in the Appendix A.

Since all concretes were performed with the same material and same concrete mix, it can be assumed that the concrete influence on the results only can be affected by the concrete own variability or the fact to be cast in different batches. A summary of the performed characterizing tests results for each batch is described in Chap. 3. Moreover, tests on the concrete long term standard properties such as creep in compression and shrinkage were also performed and described in Chap. 3. Additional information about the results of each specimen in the characterization tests can be found in the Appendix B.

A general overview description of the four main creep methodologies performed in this RRT is presented in Chap. 4. Despite this general procedure description, each participant provides a full description of the used equipment and specific details of their procedure in the Chap. 5.

Even if each laboratory performed their own methodology, the variables to be registered and the most significant test parameters of the creep test were agreed by the participants prior to start the RRT. The different data to be recorded were specified as much as possible in Chap. 6 for all the different steps in a creep test general response curve as well as the main test parameters for the RRT program. After an internal discussion, the TC reached an agreement on test parameters that represent the frame of this study like reference residual strength, pre-crack conditions, load level or test procedure details. This is an important issue since one of the objectives is to be quite representative of the working conditions in a SLS, but at the same time it was also necessary to guarantee results in a logical testing time. The data obtained for each specimen from the creep test was provided by each participant by means of an Excel sheet to be analysed. The structure of the collected data sheet is fully described in Appendix C.

The results presented in Chap. 7 comprise only a part of all the results obtained from the RRT. For each participant, a brief table of data of flexural tests as well two figures of the delayed deformations obtained in creep tests are presented. It is assumed that it is impossible to perform a complete analysis of a so big amount of data in a logical time. Therefore, it was decided to keep the RRT database [18] open for the participants groups in this RRT until December 2020 and then, guarantee an open access for the research community.

Assuming that it is impossible to do a holistic analysis of all the collected data, an initial study of the results in the RRT is done in Chap. 8. First of all, the actual parameter used are analysed in Sect. 8.1 and compared with those proposed in Chap. 6, with the idea of verifying the capabilities of the test methods, and the variability among laboratories trying to do similar test. Test parameters like residual strain or strength, load level, or environmental conditions are analysed. An initial statistical study on the influence of the different parameters in the delayed deformation, including the laboratory influence is done and exposed in Sect. 8.2. A classification of the laboratories in different clusters shows the reality of different methodological test details when using flexural test. Section 8.3 goes deeper into the analysis of those influencing factors. In order to fix or propose the way of representing the creep behaviour and define a "property parameter" a study of the viability, reproducibility, and repeatability of the creep coefficient or crack opening rate (COR) was presented in Sects. 8.4 and 8.5 respectively. Additional information about creep coefficients and COR results can be found in Appendix D and Appendix E. The influence or dependency on the presence of fibres in the cracked section is analysed in Sect. 8.6 as well as the correlation between different FRC specimens and the influence of ageing of concrete matrix are analysed in Sects. 8.7 and 8.8.

Finally, Chap. 9 presents the main conclusion of this RRT and the perspectives for future standardising and lacking aspects to be addressed in the next future.

1.4 Participants

Recognising the relevance of this topic, several research groups volunteered to engage in this RRT. Table 1.1 summarises the institutions, lead researchers and countries involved in the RRT. The table also includes the number used to refer to each laboratory throughout this document. Although initially 19 laboratories from all over the world and spanning across five continents agreed to participate in the RRT, only 16 were able to conclude the experimental work. Figure 1.1 shows the geographical distribution of participants over the world whereas Fig. 1.2 shows the location of participants from Europe.

Each participant proposed to perform a series of tests that was agreed by the Technical Committee. Table 1.2 summarises the creep test proposal by each participant

Table 1.1 Participant laboratories in the RRT

Laboratory	Participant institution	Main researcher	Country
1	Universitat Politècnica de València UPV	Pedro Serna/Aitor Llano	Spain
2	Universitat Politècnica de Catalunya UPC	Sergio Cavalaro	Spain
3	OTH Regensburg	Wolfgang Kusterle	Germany
4	BASF Construction Chemicals Italy	Sandro Moro	Italy
5	LEMIT-CIC and Facultad de Ingeniería UNLP	Raúl Zerbino	Argentina
6	Indian Institute of Technology Madras IITM	Ravindra Gettu	India
7	NV BEKAERT SA	Gerhard Vitt	Belgium
8	Tohoku University	Tomoya Nishiwaki	Japan
*9	ArcelorMittal Fibres	Sebastien Wolf	Luxembourg
10	BBRI Belgium Building Research Institute	Benoît Parmentier	Belgium
11	University of Bologna	Nicola Buratti/Claudio Mazzotti	Italy
12	Universidade Federal de Rio de Janeiro	Romildo Dias Tolêdo Filho	Brazil
13	École Polytechnique of Montréal	Jean-Philippe Charron	Canada
*14	Politecnico di Milano	Marco di Pisco	Italy
15	Sigma Béton	Damien Rogat	France
	CETU Centre for Tunnel Studies, MEDDE	Catherine Larive	France
16	Stellenbosch University	William P. Boshoff	South Africa
17	TSE Technologies in Structural Engineering	E. Stefan Bernard	Australia
18	VSH VersuchsStollen Hagerbach	Volker Wetzig/Michael Kompatscher	Switzerland
*19	IBAC Aachen	Wolfgang Brameshuber	Germany

*Laboratories that could not conclude the RRT

laboratory and the required specimens. The most extended creep procedure is the flexural creep on prismatic specimens with 14 tentative laboratories from a total of 19 initial participants. Regarding the direct tension and the square panel creep tests, there were 3 tentative participants for each methodology, whereas for flexural creep on round panels there was only one participant laboratory.

Fig. 1.1   Geographic distribution of RRT participants around the world

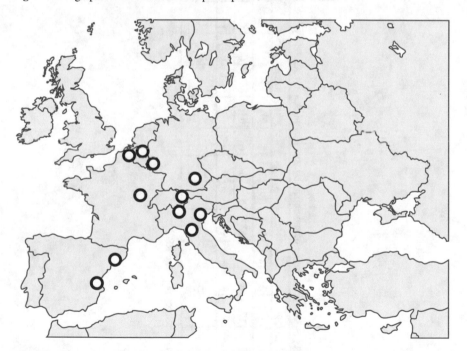

Fig. 1.2   Geographic distribution of RRT participants around Europe

Table 1.2 Creep test proposal and number of specimens by tentative participant laboratory

Creep test	Participant																			Total
	1	2	3	4	5	6	7	8	9*	10	11	12	13	14*	15	16	17	18	19*	
Flexural	12	12	6	6	12	6	6	6	6	8	6	4	2	–	–	–	–	–	6	98
Direct tension	–	–	–	–	–	–	–	–	–	–	6	–	–	6	–	4	–	–	–	16
Square panel	–	–	–	–	–	–	–	–	–	–	–	4	–	–	12	–	–	8	–	24
Round panel	–	–	–	–	–	–	–	–	–	–	–	–	–	–	–	–	4	–	–	4

*Laboratories that could not conclude the RRT

Chapter 2
Round-Robin Test Program and Execution

Aitor Llano-Torre, Pedro Serna, and Sergio H. P. Cavalaro

Abstract The round-robin test (RRT) experimental program is described in this chapter. Two different fibre-reinforced concrete (FRC) compositions were produced with macro-synthetic fibres (SyFRC) and steel fibres (SFRC). The same concrete matrix was used in both cases to avoid additional variables. The FRC matrix composition requirements were agreed by the rest of the TC members and fibres suppliers. The mechanical properties of fibres as well as the visual aspects are described. The production of specimens for each mix was performed on 2 or 3 batches of 1.5 m^3 produced in consecutive days with 1 batch per day to avoid as much as possible any variation due to production. A total of 451 FRC specimens were cast, demould, numbered and arranged in pallets to be shipped to the respective laboratories. The distribution of specimens among the participants is presented.

2.1 Materials

2.1.1 Concrete Matrix

Two fibre-reinforced concrete (FRC) compositions were produced: one with macro-synthetic fibres (SyFRC) and one with steel fibres (SFRC). In order to avoid including additional variables to the study, approximately the same concrete matrix was used

A. Llano-Torre (✉) · P. Serna
Institute of Concrete Science and Technology ICITECH, Universitat Politècnica de València (UPV), Valencia, Spain
e-mail: aillator@upv.es

P. Serna
e-mail: pserna@cst.upv.es

S. H. P. Cavalaro
School of Architecture, Building and Civil Engineering, Loughborough University, Loughborough, UK
e-mail: s.cavalaro@lboro.ac.uk

© RILEM 2021
A. Llano-Torre and P. Serna (eds.), *Round-Robin Test on Creep Behaviour in Cracked Sections of FRC: Experimental Program, Results and Database Analysis*, RILEM State-of-the-Art Reports 34, https://doi.org/10.1007/978-3-030-72736-9_2

in both cases. This matrix was designed to achieve a concrete class C30/37. The initial set of requirements for the concrete matrix composition was proposed by the chairman of the RILEM TC 261-CCF and discussed by the rest of the TC members. Fibres producers suggested changes in the composition and all proposals were compared with EN 14845-1 [19] standard for reference concretes. Table 2.1 shows the initial and final requirements approved for the round-robin test (RRT).

Ordinary Portland cement CEM I 42.5 R and water-to-cement ratio (w/c) equal to 0.5 were used following the EN 14845-1 [19]. Table 2.2 shows the final composition of the FRC deployed in the RRT. The superplasticiser content was adjusted depending on the fibre type to guarantee similar workability. Note that 10 kg per cubic meter of concrete of macro-synthetic fibres equates to 1.1% by volume, whereas a dosage of 30 kg of steel fibres per cubic meter of concrete equates to 0.38% by volume.

Figure 2.1 presents the grading curves of each aggregate individually and the grading curve of the mixed aggregates in the proportion used in the concrete, which was adjusted to approximate the ideal Bolomey curve. Three different sand types were employed in the composition to better approximate the ideal Bolomey curve.

Table 2.1 Concrete matrix requirements

Constituents/Properties	Initial	EN 14845-1	Final
Cement type	CEM II 42.5R	CEM I 42.5R	CEM I 42.5R
Cement (kg/m^3)	300–350	350	350
W/C ratio		0.55	0.5
Compressive strength f_c	30/37	25/30	30/37
Maximum aggregate size	12	16	16
Slump	S2–S3	S2	S3
Flexural residual strength $f_{R,1}$		>1.5	1.8
Flexural residual strength $f_{R,3}$		>1.0	2.5
Steel fibre content	40		30
Macro-synthetic fibre content	9		10

Table 2.2 Concrete composition in kg/m^3

Component	SyFRC	SFRC
Cement CEM I 42.5 R	350	350
Water, w/c = 0.50	175	175
Aggregate 1 (12/20)	118	118
Aggregate 2 (6/12)	591	591
Sand 1	482	482
Sand 2	482	482
Sand 3	168	168
Superplasticizer	3.75	2.50
Fibres	10	30

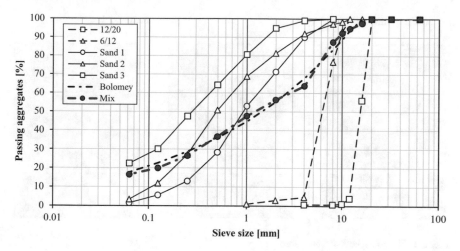

Fig. 2.1 Grading curves: individual aggregates, mixed aggregates and Bolomey

2.1.2 Fibres

Many types of fibres available nowadays can provide enough post-cracking response to be considered as structural. For the purpose of the RRT it was decided to choose as representative only one fibre and dosage for materials that are predominantly used in real applications. Considering these frame conditions, fibre type and dosage for each material were proposed by the fibre producers supporting the RRT and agreed by the TC after preliminary tests. Table 2.3 summarises the main properties of the fibres used in the RRT extracted from the product datasheet provided by the fibre suppliers.

The fibre lengths are representative of common FRC applications. Fibre dosages have been discussed and established in a way that residual strength performance is at comparable level regardless of the material of fibre. In the case of the SFRC

Table 2.3 Characteristics of fibres reported on the commercial data sheet of fibre suppliers

Properties	Macro-synthetic fibres	Steel fibres	
Supplier	BASF	BEKAERT	ArcelorMittal Fibres
Brand	–	Dramix 3D 65-60BG	HE 90/60
Material	Polypropylene	Steel	Steel
Fibre length L (mm)	40	60	60
Wire diameter d_{eq} (mm)	0.75	0.90	0.90
Aspect ratio ($\lambda = L/d_{eq}$)	53	67	67
Tensile strength (MPa)	430	1160	1200
Young's modulus (MPa)	3400	210,000	210,000

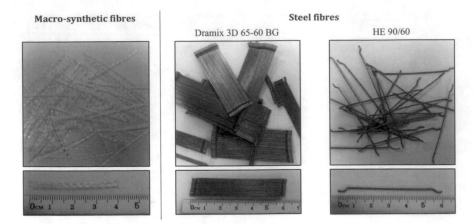

Fig. 2.2 Visual aspects of fibres used

mixes, steel fibres from both suppliers were combined and 50% by weight of Dramix 3D 65-60BG from Bekaert and 50% by weight of HE 90/60 from ArcelorMittal Fibres were mixed. Fibres of both suppliers are of comparable shape (simple end-hook) and similar fibre properties. Macro-synthetic fibres were provided from one manufacturer. They are of irregular sectional shape and are crimped in longitudinal direction. Figure 2.2 shows the visual aspect of the fibres used.

2.2   Production and Delivery

To ensure that nearly the same concrete composition and components were tested in all laboratories, specimens were produced in the same location and then, shipped to each laboratory. The concrete production for this RRT represented a major challenge due to the high number of specimens produced per day and due to the logistics behind their shipment. More than 14 tons of FRC were batched to cast 451 specimens (including prisms, cylinders, and square and round panels). The production was conducted in a precast concrete plant with an industrial mixer and casting benches. Special moulds were designed for all dimensions of the specimens to reduce the space and time of demoulding.

Initially, 0.5 m^3 batches were produced for each composition to test their fresh-state behaviour and to evaluate the casting procedure in the new moulds. After validating the concrete composition and casting procedure, 1.5 m^3 batches were produced for each composition (M-B0 for SyFRC mix and S-B0 for SFRC mix) to test the real volume and the timing in pouring the material into the moulds. Specimens cast with batch M-B0 were not used for creep tests in the RRT program, being stored in case additional tests were required.

Table 2.4 Number of specimens produced in each batch

Shape	Size (mm)	M-B1	M-B2	S-B0	S-B1	S-B2	Total
Prismatic S1	150 × 150 × 600	63	60	45	61	37	266
Prismatic S2	100 × 100 × 500	5	8	–	1	8	22
Prismatic S3	150 × 150 × 700	5	–	–	8	–	13
Square panel	600 × 600 × 100	7	9	3	7	6	32
Round panel	Ø800 × 75	2	2	–	–	–	4
Cylindrical	Ø150 × 300	30	28	6	30	20	114
Total		112	107	54	107	71	451

The production of the FRC specimens for each fibre material was performed on 2 or 3 batches of 1.5 m³ produced in consecutive days with 1 batch per day. Table 2.4 shows the number of specimens produced for each batch of FRC. Note that the letter "M" indicates batches with *macro-synthetic* fibres (SyFRC) and the letter "S" indicates batches with *steel* fibres (SFRC).

The day after production, the specimens were demoulded in the early morning, marked with the batch reference and numbered according to the casting order to identify potential differences induced by the production order (see demoulded specimens in Fig. 2.3). After that, the moulds were reassembled and placed again over the casting bench to produce the following batch.

Three different shapes of specimens were cast for creep test in cracked section as seen Fig. 2.4: prismatic, square panels and round panels. Two different size of prismatic specimens were cast: 150 × 150 × 600 mm for flexural tests and 100 × 100 × 500 for direct tension tests.

The specimens were demoulded and stored during 28 days in ambient conditions. The casting plant was located in the town of Chiva, 30 km far away from Valencia. During the storage period, the average temperature was 25.5 °C with a coefficient of variation of 5.2% and a mean relative humidity of 60.5% and 14.4% of coefficient of

Fig. 2.3 Cylindrical specimens and specimens in the pre-cast factory vibrating bench

Fig. 2.4 Demoulded FRC specimens: **a** prismatic, **b** square panels and **c** round panels

Fig. 2.5 Concrete specimens on pallets: **a** prismatic, **b** square panels and **c** round panels

variation. By the end of the storage period, the specimens were arranged in pallets as seen in Fig. 2.5 ready to be shipped to the respective laboratories.

Shipments to European laboratories were made by road. For the rest of laboratories, the shipment was made by ship or aeroplane, depending on the destination country and shipment duration. The starting point date for the creep test was established near 20th September, and all the specimens should be at the destinations time in advance enough to be able to perform the pre-cracking tests and previous preparation if needed. Depending on the laboratory and due to transport and customs issues in some countries, the age of specimens when starting the creep test may range from 90 days to 140 days. In the case of Canada, Brazil, Argentina and South Africa, the specimens were sent by maritime transport, whereas the specimens for Australia, Japan and India were sent by plane to destinations due to the long duration of the delivery time by ship. Despite this effort to reduce the transport time, the specimens of the LAB-05 participant were detained by the Argentinian customs for four months, and thus, this participant started the creep teste some months later. Table 2.5 summarises the total number of specimens sent to each destination. Note that not all the specimens sent to the participants were tested in creep. Additional specimens for characterization purposes or additional tests were also sent. The total number of FRC specimens tested in creep by participant is exposed in Table 4.1.

Table 2.5 Specimen distribution

Specimen	Participant																			Total
	1	2	3	4	5	6	7	8	9	10	11	12	13	14	15	16	17	18	19	
Prismatic S1	95	20	10	36	20	15	10	10	–	8	20	10	–	2	–	–	–	–	10	266
Prismatic S2	–	–	4	–	–	–	–	2	–	–	–	–	–	–	–	16	–	–	–	22
Prismatic S3	–	–	–	–	–	–	–	–	10	–	–	–	3	–	–	–	–	–	–	13
Square P.	–	–	–	–	–	–	–	–	–	–	–	4	–	–	20	–	–	8	–	32
Round P.	–	–	–	–	–	–	–	–	–	–	–	–	–	–	–	–	4	–	–	4
Cylindrical	80	–	–	–	4	–	–	–	–	–	30	–	–	–	–	–	–	–	–	114
Total	175	20	14	36	24	15	10	12	10	8	50	14	3	2	20	16	4	8	10	451

2.3   Schedule of RRT

The schedule of RRT spanned from March 2015 to December 2019, as follows:

- March 2015—Round-Robin Test approval
- June 2015—Specimens production
- July 2015—Specimens shipment to destinations
- September 2015—Start of creep test around the world
- September 2016—Stop creep test and start 1-month recovery
- October 2016—Completion of 1-month recovery and test to failure
- March 2017—1st discussion about results in Bologna
- September 2017—1st statistical analysis discussion in Leuven (data incomplete)
- February 2018—Database and checklist collection
- September 2018—1st Round-Robin Test Report draft
- March 2019—2nd Round-Robin Test Report draft
- 2019–2020—Round-Robin Test Report publication.

Chapter 3
Fibre Reinforced Concrete Characterization

Aitor Llano-Torre and Pedro Serna

Abstract The characterisation of the basic properties of all the fibre-reinforced concrete (FRC) batches of the round-robin test (RRT) was performed by the organising laboratory in LAB-01 facilities. The basic characterisation comprised the assessment of the compressive strength, elastic modulus and flexural tensile strength. The compressive strength and the flexural tensile strength were measured at 7 and 28 days for control purposes. Moreover, characterisation tests were performed at 3 different time lapses during the RRT to evaluate ageing influence on flexural properties evolution during the creep test duration: T1 at the beginning of the creep test, T2 at mid-term of creep test or 180 days and T3 by the end of creep test. Complementary long-term tests such as shrinkage and creep in compression were performed during the RRT for each FRC batch by LAB-01 and LAB-11. This section presents an overview of the results of the charactcrization tests and additional information such as individual results or residual behaviour charts can be found in Appendix B .

3.1 Compressive Strength and Elastic Modulus

Once demoulded, the specimens were stored when received in a moist chamber waiting to be tested. Table 3.1 shows the compressive strength and the elastic modulus assessed, respectively, according to EN 12,390–3:2009 [20] and EN 12,390–13:2013 [21] using cylindrical specimens. The compressive strength at 28 days is slightly below the expected in a C30/37. Despite that, a similar compressive strength was obtained in all batches, with values ranging from 30.8 to 37.6 MPa at 28 days. The compressive strength increases with the concrete age for all batches except for M-B2 at T2.

A. Llano-Torre (✉) · P. Serna
Institute of Concrete Science and Technology ICITECH, Universitat Politècnica de València (UPV), Valencia, Spain
e-mail: aillator@upv.es

P. Serna
e-mail: pserna@cst.upv.es

© RILEM 2021

A. Llano-Torre and P. Serna (eds.), *Round-Robin Test on Creep Behaviour in Cracked Sections of FRC: Experimental Program, Results and Database Analysis*, RILEM State-of-the-Art Reports 34, https://doi.org/10.1007/978-3-030-72736-9_3

Table 3.1 Average compressive strength and elastic modulus results

Mix	Batch	Mean f_c (N/mm²) (CV in %)					E (N/mm²)
		7 days	28 days	T1 (~90 days)	T2 (~300 days)	T3 (~500 days)	T1
SyFRC	M-B1	30.53 (1.0%)	33.93 (1.2%)	37.87 (1.1%)	40.11 (0.5%)	42.18 (2.8%)	28,633
	M-B2	31.23 (3.8%)	37.57 (0.4%)	42.50 (1.1%)	38.82 (6.5%)	44.06 (6.1%)	30,105
SFRC	S-B1	27.33 (0.6%)	30.80 (3.7%)	34.73 (1.5%)	34.45 (3.9%)	35.53 (5.1%)	–
	S-B2	30.77 (1.5%)	36.17 (1.7%)	37.90 (3.0%)	38.59 (1.1%)	–	29,631
	S-B0	31.23 (3.8%)	37.70 (0.8%)	–	–	–	–

3.2   Flexural Residual Strength

The characterisation of the residual flexural strength followed the procedure defined in EN 14651 [22].

3.2.1   Characterisation at 7 and 28 Days

Table 3.2 shows the results of flexural tests performed at 7 and 28 days. All batches present a similar limit of proportionality (LOP) values, which increase with the concrete age. Despite using nearly the same concrete composition and production

Table 3.2 Average strength at LOP and average residual flexural strength at 7 and 28 days

Mix	Batch	7 days				28 days			
		f_L	$f_{R,1}$	$f_{R,3}$	$f_{R,3}/f_{R,1}$	f_L	$f_{R,1}$	$f_{R,3}$	$f_{R,3}/f_{R,1}$
SyFRC	M-B1	3.12	1.86	3.06	1.64	3.65	2.37	3.57	1.50
	M-B2	3.29	1.93	2.96	1.54	3.45	1.99	3.08	1.54
	Average	3.20	1.90	3.01	1.59	3.53	2.15	3.28	1.52
	CV (%)	5.6	9.0	6.7	4.7	4.1	12.0	15.5	6.5
SFRC	S-B1	3.15	1.88	2.34	1.22	3.44	2.80	3.00	1.07
	S-B2	3.16	2.42	2.46	1.20	3.90	3.28	3.69	1.12
	S-B0	3.37	2.64	3.15	1.19	3.92	3.56	3.72	1.05
	Average	3.25	2.34	2.76	1.21	3.75	3.21	3.47	1.08
	CV (%)	5.9	27.9	30.2	4.3	7.0	15.4	17.0	5.9

process, a high scatter in the residual behaviour at different crack mouth opening displacement (CMOD) was observed among the batches with the same type of fibre, but especially for macro-synthetic fibre-reinforced concrete (SyFRC) mixes. Higher $f_{R,1}$ values were obtained by steel fibre-reinforced concrete (SFRC) mixes for both 7 and 28 days characterization times. Regarding the residual strength $f_{R,3}$ more comparable results for both fibre materials mixes were obtained as desired. Considering the typical scatter of this test, the results show that mixes present similar flexural response, regardless of the type of fibre used. The average $f_{R,3}$ obtained is slightly higher than expected according to Table 2.1.

Figure 3.1 contains the Stress-CMOD curves obtained in the flexural characterisation at 7 and 28 days. The higher scatter observed for SFRC can be attributed to the fibre content by volume and the consequent difference in the number of fibres crossing the cracks. Although the content by weight of macro-synthetic fibres (10 kg/m^3) is 3 times smaller than that of steel fibres (30 kg/m^3), the material of the macro-synthetic fibres has a density approximately 8 times smaller than the density of steel. This

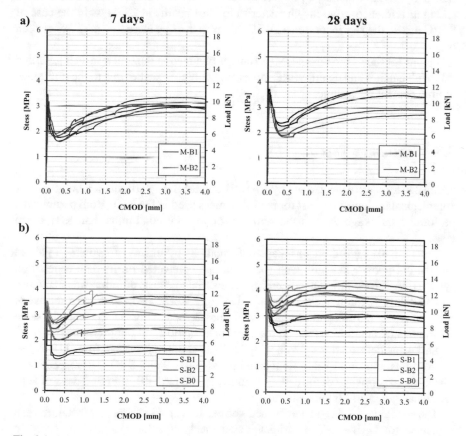

Fig. 3.1 Stress-CMOD curves at 7 and 28 days of SyFRC **a** and SFRC **b** mixes

leads to fibre content by volume almost 3 times bigger in the case of the SyFRC mixes, which contribute to reducing the scatter in the results.

3.2.2 Characterisation at T1, T2 and T3

Complementary characterisation tests were performed at 3 different time lapses during the RRT to evaluate ageing influence on flexural properties evolution during the creep test duration: T1 at the beginning of the creep test, T2 at mid-term of creep test or 180 days and T3 by the end of creep test. Table 3.3 shows the residual flexural strength assessed at T1, T2 and T3 time lapses. As happened with the matrix compressive strength, the strength at LOP (f_L) increases in time with age, going from 3.7 MPa at 28 days to values near 5 MPa at T2 and T3. The residual flexural strengths $f_{R,1}$ and $f_{R,3}$ due to the fibre reinforcement also increases with age. In the case of SyFRC mixes, the $f_{R,1}$ increased a 39.5% whereas the $f_{R,3}$ increased a 45.1% taking as reference value the characterisation test results at 28 days. In the case of SFRC mixes, the $f_{R,1}$ increased a 20.2% whereas the $f_{R,3}$ increased a 22.5% as seen on Fig. 3.2.

Figure 3.3 shows the Stress-CMOD curves obtained in flexural tests at T1, T2 and T3 for all batches and again, SFRC mixes, shows higher scatter due to the difference in fibre content by volume.

3.3 Shrinkage Test

Shrinkage tests were carried out by LAB-01 and LAB-11 participant laboratories as complementary long-term test for the FRC mixes used in the RRT. Both participants performed shrinkage tests in the same environmental conditions than RRT creep tests.

The LAB-01 registered shrinkage deformations by means of two strain gauges located in opposite sides of each specimen and connected to the data acquisition system (DAS) for continuous recording. The shrinkage tests duration was 360 days, and the delayed evolution can be observed in Fig. 3.4. The LAB-01 located shrinkage specimens over the creep in compression frames as seen in Fig. 3.6 to assure the same environmental conditions than all the specimens involved in long-term test.

Shrinkage deformations obtained by LAB-01 at 360 days range from 118 to 189 microstrains, threshold values registered by M-B1-019 and S-B1-240 respectively. The average shrinkage deformation obtained by the LAB-01 after 360 days of tests for all batches is 142.6 microstrains.

The participant LAB-11 performed shrinkage test with similar procedure than explained for LAB-01. The shrinkage specimens were located near the creep in compression frames in similar climate conditions that the creep in compression specimens. Two SyFRC and three SFRC specimens were tested in shrinkage. The

Table 3.3 Average strength at LOP and average residual flexural strengths at T1, T2 and T3

Mix	Batch	T1 (0 days in creep)				T2 (180 days in creep)				T3 (360 days in creep)			
		f_L	$f_{R,1}$	$f_{R,3}$	$f_{R,3}/f_{R,1}$	f_L	$f_{R,1}$	$f_{R,3}$	$f_{R,3}/f_{R,1}$	f_L	$f_{R,1}$	$f_{R,3}$	$f_{R,3}/f_{R,1}$
SyFRC	M-B1	3.98	2.55	3.94	1.55	5.11	2.95	4.87	1.65	4.76	2.84	4.53	1.60
	M-B2	3.92	1.92	2.97	1.55	5.45	3.06	4.64	1.51	4.96	2.42	3.63	1.51
	Average	3.95	2.23	3.46	1.55	5.28	3.00	4.76	1.58	4.86	2.63	4.08	1.55
	CV (%)	1.1	19.5	19.8	1.3	4.5	10.9	15.5	6.8	7.5	9.9	13.6	8.3
SFRC	S-B1	3.82	2.77	2.77	0.99	4.56	3.35	3.79	1.13	4.27	3.85	4.10	1.05
	S-B2	3.90	3.79	4.38	1.15	4.84	3.47	3.69	1.07	4.99	3.86	4.39	1.14
	S-B0	3.99	3.52	3.98	1.13	4.78	3.78	4.01	1.07	-	-	-	-
	Average	3.90	3.36	3.71	1.09	4.72	3.50	3.81	1.09	4.63	3.86	4.25	1.10
	CV (%)	5.4	20.3	25.9	8.5	3.8	8.9	8.9	7.7	11.6	14.8	20.2	10.1

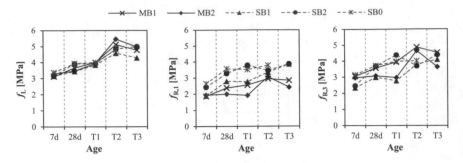

Fig. 3.2 Evolution in time of average flexural test parameters: f_L, $f_{R,1}$ and $f_{R,3}$

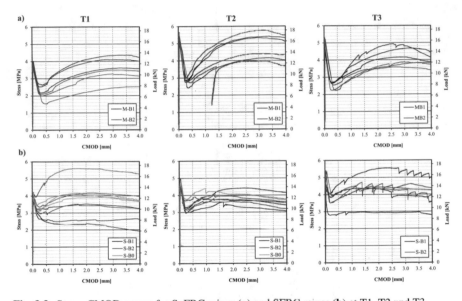

Fig. 3.3 Stress-CMOD curves for SyFRC mixes (**a**) and SFRC mixes (**b**) at T1, T2 and T3

shrinkage curves obtained from LAB-11 are depicted in Fig. 3.5. After 360 days of shrinkage test, the delayed deformations range from 118.8 to 154.4 microstrains with an average value of 134.7 microstrains. The shrinkage values obtained from both participants are quite consistent and similar.

3.4 Creep in Compression Test

LAB-01 and LAB-11 participant laboratories also carried out creep in compression test for each FRC mix in the same environmental conditions than RRT creep tests.

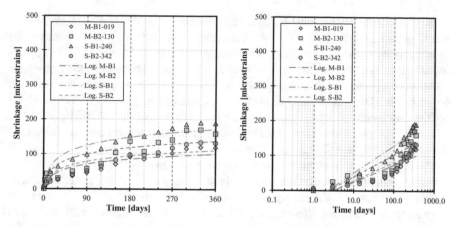

Fig. 3.4 Delayed deformations obtained from shrinkage creep test by LAB-01

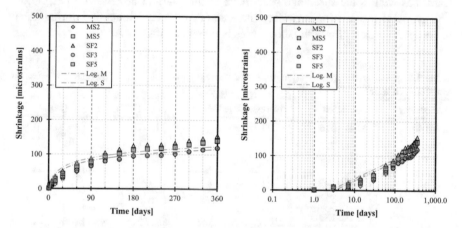

Fig. 3.5 Delayed deformations obtained from shrinkage creep test by LAB-11

Fig. 3.6 Creep in compression test frames and Shrinkage specimens: **a** LAB-01 and **b** LAB-11

Regarding the creep in compression tests performed by the LAB-01, three cylindrical specimens (Ø150 × 300 mm) from four of the batches were selected to be tested following the methodology described in the standard ASTM C512 [23]. As explained in Sect. 6.2.4, the applied load level or creep index in creep frames for creep in compression test was accorded as the 40% of mean f_c at 28 days of each tested batch. Therefore, the applied loads in the frames range from 12.35 to 14.81 kN for the four batches. Figure 3.6a shows the creep frames of the LAB-01 participant with the FRC specimens during creep in compression tests and the specimens for shrinkage measurement over each frame.

Additional creep in compression tests were performed by the LAB-11 following similar procedure but with different criteria in the applied load selection. Instead of applying a percentage of the compressive strength of each batch, the LAB-11 obtained the average compressive strength f_c at 28 days for all the batches (35.21 MPa) and then applied a stress of 7.0 MPa, the 20% of the average f_c. The creep in compression frames as well as the shrinkage specimens of the LAB-11 are shown in Fig. 3.6b. The available information regarding the load and creep index applied during creep in compression tests for both laboratories is given in Table 3.4.

Table 3.4 Average compressive strength, Applied stress and creep index of creep in compression tests for each series

Laboratory	Specimen	Average $f_{c, 28}$ (MPa)	$f_{R,c}$ (MPa)	I_c (MPa)
LAB-01	M-B1-016	33.93	13.25	39.07%
	M-B1-017			
	M-B1-018			
	M-B2-128	37.57	14.81	39.42%
	M-B2-129			
	M-B2-132			
	S-B1-237	30.80	12.35	40.09%
	S-B1-238			
	S-B1-239			
	S-B2-339	36.17	13.63	37.68%
	S-B2-340			
	S-B2-341			
LAB-11	M-B1	33.93	7.0	20.63%
	M-B2	37.57	7.0	18.63%
	S-B1	30.80	7.0	22.73%
	S-B2	36.17	7.0	19.35%

3.4.1   Compressive Creep Test Delayed Deformations

After one year of loading, the delayed strains curves of each batch obtained by LAB-01 and LAB-11 are depicted in Figs. 3.7 and 3.8 respectively. The average batch curves in Fig. 3.7 were obtained from three specimens each. Note that for both LAB-01 and LAB-11 delayed deformations results figures, the shrinkage deformations were removed to show only delayed deformations given to the applied load during the creep in compression tests.

The delayed deformations obtained for both participants are consistent and robust with a slightly higher scatter registered by LAB-01. As it can be seen in Fig. 3.7, the delayed compressive strains of SyFRC registered by LAB-01 seem to be higher than for SFRC. In absolute values, the average deformation at 360 days obtained by

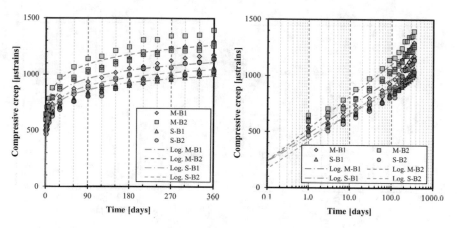

Fig. 3.7   Delayed deformations obtained from compressive creep test by LAB-01

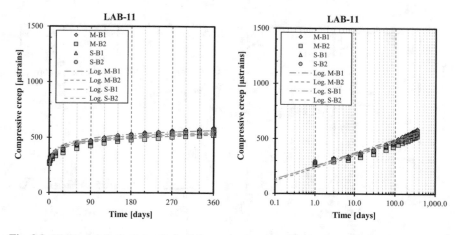

Fig. 3.8   Delayed deformations obtained from compressive creep test by LAB-11

LAB-01 is 1151.6 microns whereas the LAB-11 obtained 553.5 microns. The ratio of delayed deformations between both laboratories corresponds to the ratio between the applied creep indexes by the participants as it can be expected.

3.4.2 Compressive Creep Test Parameters Analysis

The creep coefficient and Compressive Rate (CR) parameters were calculated to compare the results obtained in compressive creep test performed by the laboratories. Therefore, the evolution in time of the creep coefficient curves as well as the CR obtained by LAB-01 and LAB-11 are depicted in Figs. 3.9 and 3.10 respectively.

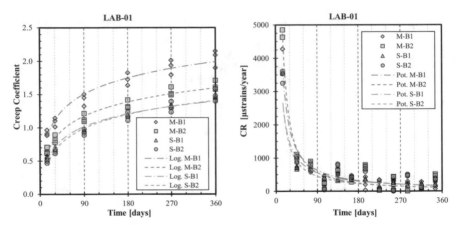

Fig. 3.9 Creep coefficients and CR calculated from compressive creep test by LAB-01

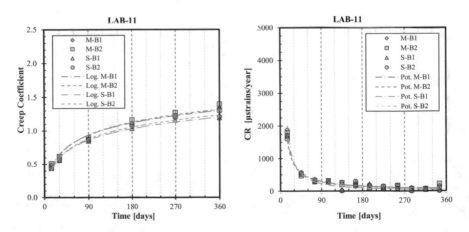

Fig. 3.10 Creep coefficients and CR calculated from compressive creep test by LAB-11

As occurs in the delayed compressive strains comparison, the creep coefficients obtained by LAB-01 are higher scattered than for the LAB-11. On the contrary, the creep coefficient ratio between both participants increases from 0.48 to 0.81 due to the influence on the instantaneous deformations at early ages, where the average creep coefficients at 360 days are 1.36 and 1.09 obtained by LAB-01 and LAB-11 respectively.

Regarding the Compressive Rate, the velocity developed by specimens of LAB-01 at 40% of f_{ck} is higher than the specimens of LAB-11 at 20% f_{ck} and the ratio is again related to the creep index ratio.

Chapter 4
General Procedure of Creep Test

Aitor Llano-Torre, Pedro Serna, William P. Boshoff, Nicola Buratti, Karyne F. dos Santos, and E. Stefan Bernard

Abstract Regardless of the methodology adopted by each laboratory, there is a consensus about the three main stages in the creep testing procedure: the pre-cracking stage, the creep stage and the post-creep stage. Despite this general procedure, four main methodologies were conducted in fibre-reinforced concrete (FRC) specimens in the round-robin test (RRT): flexure creep test on prismatic specimens based on EN 14651, direct tension creep test, flexure creep test on square panels based on EN 14488-5 and flexure creep test on round panels based on ASTM C1550. An overview of these four main methodologies is described in the following subsections. All the procedures defined by each participant were grouped by methodology and compared in terms of configuration, equipment and measurement devices.

A. Llano-Torre (✉) · P. Serna
Institute of Concrete Science and Technology ICITECH, Universitat Politècnica de València (UPV), Valencia, Spain
e-mail: aillator@upv.es

P. Serna
e-mail: pserna@cst.upv.es

W. P. Boshoff
Faculty of Engineering, Built Environment and Information Technology, University of Pretoria, Pretoria, South Africa
e-mail: billy.boshoff@up.ac.za

N. Buratti
Department of Civil, Chemical, Environmental and Materials Engineering DICAM, University of Bologna, Bologna, Italy
e-mail: nicola.buratti@unibo.it

K. F. dos Santos
Department of Civil and Environmental Engineering, University of Brasília, Brasília, Brazil
e-mail: karyne.ferreira@aluno.unb.br

E. S. Bernard
TSE Technologies in Structural Engineering Pty Ltd, Sydney, Australia
e-mail: s.bernard@tse.net.au

© RILEM 2021
A. Llano-Torre and P. Serna (eds.), *Round-Robin Test on Creep Behaviour in Cracked Sections of FRC: Experimental Program, Results and Database Analysis*, RILEM State-of-the-Art Reports 34, https://doi.org/10.1007/978-3-030-72736-9_4

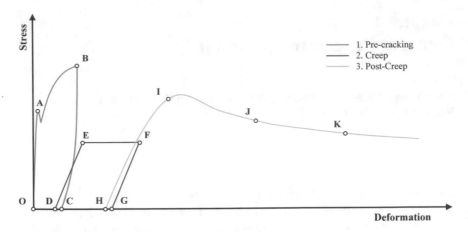

Fig. 4.1 Complete diagram of main stages of a creep test

An idealized curve of the complete creep test procedure is illustrated in Fig. 4.1. This diagram is valid for creep tests where the measured deformation is crack mouth opening displacement (CMOD), crack opening displacement (COD) or deflection (δ). Regardless of the methodology adopted by each laboratory, there is a consensus around the main stages in the testing procedure, which contains three main stages. In the *pre-cracking stage* (stretch OD in Fig. 4.1), the specimens are tested up to the desired crack opening and then, unloaded. In the *creep stage* (stretch DH in red in Fig. 4.1), the specimens are reloaded up to a specific stress (expressed as a percentage of the residual strength capacity) that is maintained for a particular time and then unloaded. Finally, during the *post-creep stage* (stretch HK in Fig. 4.1), the specimens are tested until failure after the creep test.

Despite this general procedure, the four main methodologies must be also well defined: *flexural creep* on prismatic specimens or beams, *direct tension* creep, flexural creep on *square panels* and flexural creep on *round panels*. An overview of these four main methodologies is described in the following subsections. Additional information about the specific procedure followed by each laboratory is presented in Chap. 5.

Table 4.1 summarises the creep test conducted in the round-robin test (RRT) by each participant laboratory. The most extended method is the flexural creep for prismatic specimens with 12 out of 16 participant laboratories. Direct tension creep tests were only performed by 2 participants. Flexural creep tests of square panels were performed in 3 laboratories, whereas the flexural creep of round panels was only conducted in one laboratory.

Table 4.1 Creep test methodology and number of specimens tested in each laboratory

Creep test	Participant laboratory																			Total
	1	2	3	4	5	6	7	8	9*	10	11	12	13	14*	15	16	17	18	19*	
Flexural	12	12	6	6	12	6	6	6	•	8	6	4	2	–	–	–	–	–	•	86
Direct tension	–	–	–	–	–	–	–	–	–	–	6	–	–	•	–	4	–	–	–	10
Square panel	–	–	–	–	–	–	–	–	–	–	–	4	–	–	12	–	–	8	–	24
Round panel	–	–	–	–	–	–	–	–	–	–	–	–	–	–	–	–	4	–	–	4

*Laboratories that could not conclude the RRT

4.1 Flexural Creep Tests Procedure on Prismatic Specimens

Several test configurations for the *flexural creep test* can be found in the literature. Participant laboratories in the RRT have developed or adapted their methodology and the main characteristics are summarised in Table 4.2. Some of the differences between laboratories are given by the national reference standard available to characterise the concrete flexural strength, such as the existence or absence of a notch in the specimens and the three-point bending test (3PBT) or four-point bending test (4PBT) load configuration during the tests. Moreover, the creep frames design, the load transfer system or the control of the environmental conditions are additional differences to be considered.

Considering the significant number of laboratories in the RRT and the goal of assessing the influence of the methodology in the creep results, the TC decided to allow each laboratory to maintain their respective procedure. The analysis of the results should detect the significance of those differences on the methodology. This section presents a general description of the *Flexural Creep Test Procedure*. Information about specific procedure and equipment used by each laboratory can be found in Chap. 5.

Table 4.2 Main differences in flexural creep methodologies performed by the laboratories in the RRT

Participant	Notch	Load configuration		Transducer	Measure	Multi-specimen	Load transfer	Clim. control
		Pre-crack	Creep					
LAB-01	Yes	3 PBT	4 PBT	Electronic	CMOD	Yes (×3)	Lever arm	Yes
LAB-02	Yes	3 PBT	4 PBT	Analogic	CMOD	Yes (×3)	Lever arm	No
LAB-03	No	4 PBT	4 PBT	Analogic	δ	No	Lever arm	No (wrapped)
LAB-04	Yes	3 PBT	4 PBT	Electronic	CMOD	Yes (×3)	Lever arm	Yes
LAB-05	Yes	3 PBT	4 PBT	Electronic	CMOD	Yes (×3)	Lever arm	Yes
LAB-06	Yes	3 PBT	4 PBT	Electronic	CMOD	Yes (×3)	Lever arm	Yes
LAB-07	Yes	3 PBT	3 PBT	Analogic	δ	No	Lever arm	No
LAB-08	Yes	4 PBT	4 PBT	Analogic	CMOD	Yes (×2)	Screw bars	Yes
LAB-10	Yes	3 PBT	3 PBT	Electronic	CMOD	No	Hydraulic jack	Yes
LAB-11	Yes	3 PBT	4 PBT	Electronic	CMOD	Yes (×3)	Lever arm	Yes
LAB-12	Yes	4 PBT	4 PBT	Electronic	CMOD	Yes (×2)	Hydraulic jack	Yes
LAB-13	Yes	4 PBT	4 PBT	Electronic	CMOD/δ	Yes (×2)	Hydraulic jack	Yes

4.1.1 Pre-cracking Stage

Most participants performed the pre-cracking test following the EN 14651+A1:2007 standard [22] with notched specimens tested in a 3PBT load configuration. Some laboratories adopted the OBV 2008 guideline [24] or German guideline for steel fibre-reinforced concrete (SFRC) [25] that recommend the use of un-notched specimens tested in a 4PBT load configuration. Figure 4.2 shows different load configurations adopted for pre-cracking tests by participants.

In all laboratories, the pre-cracking was performed with CMOD or deflection control so that the piston of the press ensures a constant CMOD rate of 0.05 mm/min up to the desired CMOD of 0.5 mm or the equivalent deflection (Point B Fig. 4.1). This rate is low enough to allow a clear definition of the limit of proportionality (LOP) and the first residual strength ($f_{R,1}$) of the specimens. The maximum CMOD or deflection reached was recorded, and the specimen was unloaded at a CMOD rate of 2 mm/min. The CMOD or deflection elastic recovery was recorded once the load was removed (Point C in Fig. 4.1) and 10 min later a delayed recovery was also recorded (Point D in Fig. 4.1). Then, the specimens were removed from the press, turned 90° to avoid unintended crack openings due to the dead load or the transport and stored until the beginning of the creep stage.

Participants were advised to store the specimens in similar environmental conditions than in the location of the creep frames. For instance, if the creep frames were in a room with climate-controlled conditions, the specimens should have been stored inside this climate-controlled room to ensure the specimens would reach similar temperature and humidity, thus reducing variations in crack opening induced by temperature or humidity changes.

 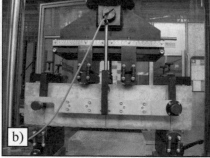

Fig. 4.2 Different load configurations in pre-cracking stage: **a** 3PBT in notched specimens, **b** 4PBT in un-notched specimens measuring deflection.

4.1.2 Creep Stage

The specimens can be tested individually or in a multiple setup formed by two or more specimens. In the latter case, 4PBT setup was usually adopted to improve the stability of the column. The change in load configuration between pre-cracking (3PBT) and creep (4PBT) stages implied a numerical transformation of the load reached in the pre-cracking stage and used to estimate the load level applied in the creep stage. Different examples of single or multiple specimen setups are depicted in Fig. 4.3.

In the case of specimens tested individually, each specimen was placed in the creep frame, and the initial measurement of electronic or manual transducers was recorded. The load transfer mechanism was prepared, and the creep frame gradually loaded up to the desired load. In the case of the multi-specimen setup, the mounting was performed one specimen at a time. The first specimen was placed in the frame, and the measurement of the displacement measured by the transducers was recorded. Then, the second specimen was placed above the first following the same procedure, which was also repeated for the third one. Finally, the load cell and the load transmission element located at the top part of the frame were gently placed over the last specimen. Following this procedure, the crack opening variations due to the weight of the load

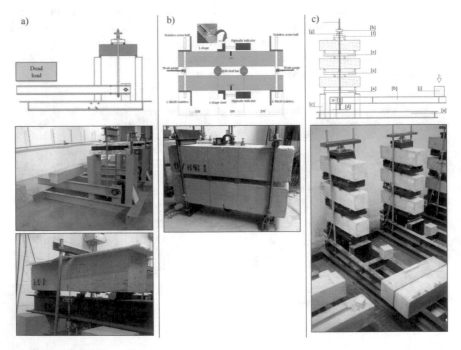

Fig. 4.3 Different specimen setup used in flexural creep methodologies: **a** single specimen setup (LAB-03, LAB-10), **b** double specimen setup (LAB-08) and **c** multi-specimen setup (LAB-01)

transmission element and the upper specimens were recorded. After that, the load was applied to the creep frame marking the start of the creep stage (Point E in Fig. 4.1). Depending on the laboratory procedure and equipment, this load was introduced by placing a dead load over the lever arm, by hydraulic jacks or by tightening screwed bars and nut.

After 360 days under sustained load (Point F in Fig. 4.1), the specimens were unloaded (Point G in Fig. 4.1) and remained in the creep frames with the same configuration to avoid unexpected sudden deformations if moved. Only in those cases where configuration with multiple specimens became unstable without load, the upper specimen could be carefully removed and placed near the frame to ensure similar environmental conditions. Recovery deformations were recorded for 30 days after unloading (until Point H in Fig. 4.1) at the same time spans as at the beginning of the creep stage.

4.1.3 Post-creep Failure Test Stage

In the post-creep stage, specimens were tested to failure to identify the influence of the delayed deformations in the residual behaviour of the FRC specimens. The post-creep failure test followed a similar procedure than that followed in the pre-cracking stage. The specimens were placed in the testing machine carefully to avoid sudden blows. If possible, participants would perform one *hysteresis loop* until the specific load applied during the creep test was reached and then, continue loading up to failure, as represented in Fig. 4.4. The post-creep tests were performed in CMOD or deflection control at a constant CMOD rate of 0.2 mm/min up to a CMOD of 4 mm or the equivalent for deflection.

Once the specimens failed, the fibres of the crack section were counted to provide additional information for the analysis of the results. In the case of prismatic specimens tested in bending, it was accorded to divide the cracked section into 3 parts

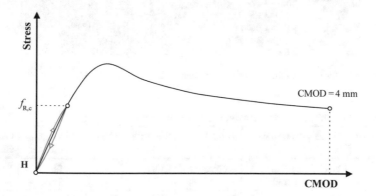

Fig. 4.4 Post-creep bending test with the initial hysteresis loop

(bottom, middle, top) with similar dimensions and the fibres counted at both sides of the crack. More information about fibre counting can be found in Sect. 8.6.

4.2   Direct Tension Creep Tests Procedure

There are two laboratories that performed the *direct tension* creep test. Although the creep load application was similar between the two laboratories, the specimen preparation differed significantly from one laboratory to the other. LAB-11 drilled cylindrical cores sized Ø94 × 150 mm from 150 × 150 × 600 mm prisms and then glued plates with threaded holes to apply the load (see Fig. 4.5a) while LAB-16 used 100 × 100 × 500 mm cast prisms and glued hooks to apply the load (see Fig. 4.5b). The former tested six specimens, whereas the latter tested four specimens. All of them were notched before testing commenced (8 mm circumferential notch for LAB-11 and 10 mm around all sides for LAB-16). The pre-cracking and creep test setups are explained in the following two sections. Table 4.3 summarises the main methodological aspects adopted by both laboratories.

Fig. 4.5 Setup for the pre-cracking of the samples: **a** LAB-11 and **b** LAB-16

Table 4.3 Main differences between direct tension creep methodologies

Laboratory	Notch	Specimen shape	Transducer	Measure	Multi	Load transfer	Climate control
LAB-11	Yes	Cylindrical (cored)	Electronic	COD	Yes (×3)	Lever arm	Yes
LAB-16	Yes	Prismatic (moulded)	Electronic	COD	No	Lever arm	Yes

4.2.1   Pre-cracking Stage

The test setup adopted for the pre-cracking stage of the samples by both participants is shown in Fig. 4.5. LAB-11 pre-cracked specimens in a servo-hydraulic actuator with closed-loop capabilities. Spherical joints were connected to the top and bottom of each specimen to simulate the same kind of boundary conditions applied in the creep test. Three clip-on gauges were connected equally spaced around the specimen crossing the notch (see Fig. 4.5a). The pre-cracking test displacement speed was controlled by the maximum clip-on gauge measurement at a rate of 0.005 mm/min. The pre-cracking of the samples in LAB-16 was also done in a servo-hydraulic actuator. In this case, the machine was controlled using one of the two LVDT symmetrically distributed and crossing the notch (see Fig. 4.5b), which led to a stable crack grown. A displacement rate of 0.05 mm/min was used. The boundary conditions were not completely free to rotate, but the system had a low rotations stiffness.

4.2.2   Creep Stage

The *direct tension* creep tests were conducted in creep frames that may allow test one specimen or multiple specimens at once if needed. LAB-11 tested three samples in series whereas LAB-16 tested one sample by creep frame. Figure 4.6 shows the setups for the two laboratories.

LAB-11 loaded the specimens by slowly adding weights to the lever arm throughout 30 s. The crack opening was measured using three LVDTs symmetrically

Fig. 4.6 Tensile creep frame used by **a** LAB-11 and **b** LAB-16

spaced around the specimen. A NI-SCXI system was used for the measurement at a rate of one reading per hour (during the first day every 5 min). The boundary conditions of the specimens were free to rotate. The temperature was controlled at 21 °C, and the relative humidity was 55%.

LAB-16 used a setup with two equal lever arms and a screw stopper at the top, which was loosened carefully to load the specimens. The load was applied over a time ranging from 38 to 54 s. The LVDT readings were taken at least every 6 h, but at a higher rate in the beginning. A HBM Spider8 setup was used for data acquisition. The temperature was controlled at 25 °C, and the relative humidity was controlled at 65%.

4.2.3 Post-creep Failure Test Stage

LAB-11 performed the post creep test up to failure using the same setup as for the pre-cracking stage with a displacement rate of 0.1 mm/min in terms of COD. However, LAB-16 used a different setup from that deployed for the pre-cracking stage. As shown in Fig. 4.7, the tests were done in a servo-mechanical testing machine (Zwick Z250) without closed-loop control and a crosshead displacement of 0.2 mm/min.

Fig. 4.7 The tensile test setup used by LAB-16 for the post creep tensile test

After failure, fibres were counted by LAB-16 and their distribution analysed on both sides of the cracking plane.

4.3 Square Panel Creep Tests Procedure

The Square Panel Creep test was performed by three participant laboratories of the RRT: University of Rio de Janeiro—UFRJ (LAB-12), Sigma Béton laboratory in association with the Centre for Tunnel Studies (CETU) of France (LAB-15) and VersuchsStollen Hagerbach laboratory in Switzerland (LAB-18).

The LAB-12 participant performed creep test on square panel following similar methodology than for flexure creep test and using a multi-specimen setup. On the other hand, both LAB-15 and LAB-18 followed similar methodologies but with some differences. The LAB-18 followed the creep test procedure recommended by EFNARC [26], whereas the LAB-15 performed creep test following the specific procedure defined for the ASQUAPRO program [27], similar to the EFNARC recommendation. The creep test procedure was specifically designed to understand the use of fibre-reinforced shotcrete (FRS) for tunnel applications. In such, the concrete is submitted to a statically indeterminate condition because of the surrounding rock support. In all cases, and in order to understand the creep phenomena in the FRS, the tests were developed based on the EN 14488-5 standard [28].

The differences between the different methodologies and procedures are presented in Table 4.4. The creep test frame construction appears as the main difference. Meanwhile both LAB-15 and LAB-18 used a single specimen creep test setup and the load was applied by a lever arm with counterweights, the LAB-12 performed a multi-specimen setup where the load was induced by means of a hydraulic jack instead. Additional information about the methodologies of each laboratory can be found in Chap. 5.

Table 4.4 Main differences between flexural panel creep methodologies

Laboratory	Specimen shape	Transducer	Measure	Pre-crack	Multi	Load transfer	Climate control
LAB-12	Square panel	Electronic	δ	2 mm	Yes ($\times 2$)	Hydraulic jack	Yes
LAB-15	Square panel	Electronic	δ	LOP	No	Lever arm	No
LAB-18	Square panel	Electronic	δ	3 mm	No	Lever arm	Yes
LAB-17	Round panel	Electronic	δ	2 mm	No	Dead load	Yes

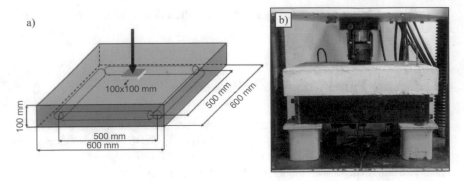

Fig. 4.8 Geometry of the pre-cracking test following the EN 14488-5 standard and view of the ongoing pre-cracking test

4.3.1 Pre-cracking Stage

The pre-cracking stage was performed by all participants following the UNE EN 14488-5 standard [28] recommendations. Each specimen was placed in a frame with a rigid square support 20 ± 1 mm thick and 500×500 mm internal dimension. The slab was adjusted in a way that the load was applied in the sprayed face. The load was transferred by a rigid steel square of 100×100 mm and thickness of 20 ± 1 mm centred in the specimen, between the specimen and the square support was laid out a layer of rubber. As for the other side, the smooth side of the test, to measure the displacement (δ), linear variable displacement transducers (LVDT) were fixed in the centre, as illustrated in Fig. 4.8.

Afterwards, the tests were carried out in a displacement control test machine at a constant rate of 1 ± 0.1 mm/min at the centre of the slab. During the pre-cracking test, the load at LOP (F_L) and the desired pre-crack deflection ($F_{R,p}$) of the specimens were obtained. When LOP or the target deflection (δ_p) was reached, depending on the laboratory procedure, the test was stopped and then the specimen unloaded. However, the transducers continued to register the data for ten minutes more. The final data was the delayed displacement recovery (Point D). Once the pre-cracking tests were done, the specimens were moved to the creep room for acclimatization when available, where the climate and humidity were controlled. By doing so, the temperature and humidity did not interfere in the crack opening variation.

4.3.2 Creep Stage

As recently explained, there were two different types of creep rigs for square panel creep tests. The UFRJ laboratory (LAB-12) tested two specimens in symmetry per creep rig using a hydraulic jack, whereas both the Sigma Béton (LAB-15) and

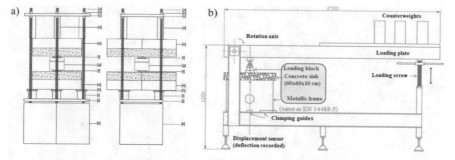

Fig. 4.9 **a** Double specimen set up using hydraulic jack by LAB-12; **b** single specimen set up using a lever arm by LAB-15 and LAB-18

the VHS laboratories (LAB-18) tested one specimen per set up where the load was applied by a lever arm and counterweights. Both square panel creep frames construction is exposed in Fig. 4.9.

Once all specimens were pre-cracked individually and positioned in the creep device, in accordance with the standard EN 14488-5 [28], a linear variable transducer was installed in the centre of the specimen, but away from the crack. In the case of the LAB-12, it was also placed two extra transducers to measure the crack opening in both directions. Before the load transferring, the transducers were set to zero. The load was transferred gradually to the established loading level by the different participant laboratories: 60% of the F_L for LAB-12, 120% of the F_L for LAB-15 and 50% of the $F_{R,p}$ for LAB-18. In addition, during the load transfer, the instantaneous deflection was registered. To have an accurate creep curve, the readings were measured not less than one per hour by a data taker. The LAB-12 controlled the load during the creep test by means of a pressure gauge, and if the load was decreasing over time due to creep deformations, the load was readjusted. On the other hand, in LAB-15 and LAB-18 the load was calibrated prior to the creep test by means of a load cell and checked again at the end of the creep test.

After the 360 days, the slabs were unloaded, but not removed from the creep rigs. In the case of the LAB-12, the load was set to 1% of the load to maintain the stability of the creep rig. The data was recorded for 30 days more in order to measure the recovery delayed deformations.

4.3.3 Post-creep Failure Test Stage

Subsequently the 30 days of recording the recovery delayed deformations, the panel creep test finished. The specimens were moved out of the creep rigs and transferred to a universal testing machine to start the post failure test stage. Each slab was tested until failure following the EN 14488-5 [28] standard, the procedure is the same as the pre-cracking stage. The specimen, as illustrated in Fig. 4.10, usually cracked in

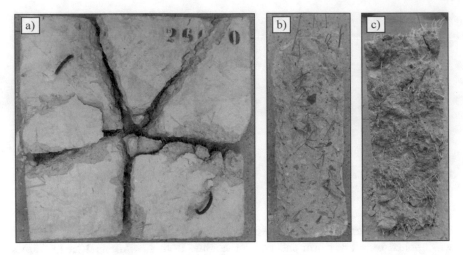

Fig. 4.10 **a** SFRC panel specimen after failure test; **b** close-up of SFRC specimen fibre distribution; **c** close-up of SyFRC specimen fibre distribution

four to five parts, for every crack face the number of fibres was counted by LAB-12.

4.4 Round Panel Creep Test Procedure

The creep test based on round panels were performed by LAB-17 at the TSE Pty Ltd laboratory in Sydney following a similar procedure to the flexural beam tests performed by other laboratories. Specific aspects of the Round Panel Test Procedure are described below to clarify the methodology adopted.

4.4.1 Pre-cracking Stage

The pre-cracking was performed with load frames used for the round panel test according to the ASTM C1550 [29] in an MTS Flextest GT servo-controller driving a 100 kN MTS 244 hydraulic actuator. These tests were performed in displacement control so that the loading piston advanced at a constant rate. During the initial cracking test, this rate was 2 mm/min in order to provide sufficient time to stop the piston at the desired post-crack deflection of 2.0 mm. The maximum initial deflection sustained in this test was noted, then the motion of the piston was reversed at a rate of 2 mm/min and the final central deflection of the specimen noted.

4.4.2   Creep Stage

This stage is performed in purpose-built loading devices of the type shown in Fig. 4.11. The dimensions of the supporting fixtures were the same as a conventional ASTM C1550 [29] test configuration, but the load is applied as a gravity load consisting of steel weights rather than through hydraulic action. Each of the four test machines included a heavy steel frame to achieve high load-strain stiffness, a floating gravity load assembly with inter-changeable weights to allow the load applied in each test to be varied, and a central hydraulic actuator to raise and lower the gravity load assembly during insertion of the specimen. The central displacement of the specimen was measured using an LVDT positioned above the panel to assess the motion of the piston relative to the load frame. The load was measured using a calibrated DC load-cell screwed onto the end of the piston. The creep tests were performed over 100 days inside a climate-controlled room in which the temperature was maintained at 22.9 ± 0.5 °C, and the relative humidity was controlled at $50 \pm 2\%$.

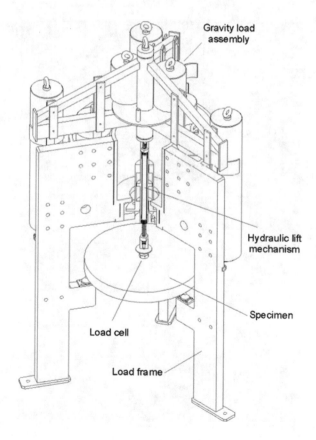

Fig. 4.11 Sectional view of the test rig used for creep gravity-load tests of round panels used by LAB-17

4.4.3   Post-creep Failure Test Stage

Upon completing the creep stage, an LVDT was placed against the underside of each panel while it still rested in the load frame depicted in Fig. 4.11. Then, the creep load was removed, and the retreat in the deflection of the centre of the panel was measured over a period of 2 h. Following completion of the unloading stage, each specimen was removed from the creep rig, transferred back to the servo-controlled testing rig, and subjected to a final ASTM C1550 [29] panel test performed at the prescribed rate of 4 mm/min piston advance to reveal the post-crack load resistance out to an additional 40 mm central deflection. This part of the testing was undertaken in the main laboratory, which was not climate-controlled but nevertheless experienced a relatively stable temperature of about 20–25 °C and relative humidity of about 60%. Given the uncertainty around the fibre counting procedure for round and square panels, the position and number of fibres in the failure surface were not assessed in these cases.

Chapter 5
Equipment and Procedure Description

Aitor Llano-Torre, Sergio H. P. Cavalaro, Wolfgang Kusterle, Sandro Moro,
Raúl L. Zerbino, Ravindra Gettu, Hans Pauwels, Tomoya Nishiwaki,
Benoît Parmentier, Nicola Buratti, Romildo D. Toledo Filho,
Jean-Philippe Charron, Catherine Larive, William P. Boshoff,
E. Stefan Bernard, and Michael Kompatscher

Abstract This chapter collects the description of the main creep testing procedures in fibre-reinforced concrete (FRC) as well as the equipment used by each participant laboratory in the round-robin test (RRT). The description of each procedure

A. Llano-Torre (✉)
Institute of Concrete Science and Technology ICITECH, Universitat Politècnica de València
(UPV), Valencia, Spain
e-mail: aillator@upv.es

S. H. P. Cavalaro
School of Architecture, Building and Civil Engineering, Loughborough University,
Loughborough, UK
e-mail: s.cavalaro@lboro.ac.uk

W. Kusterle
OTH Regensburg, Regensburg University of Applied Sciences, Regensburg, Germany
e-mail: wolfgang@kusterle.net

S. Moro
Master Builders Solutions, MBCC Group, Treviso, Italy
e-mail: sandro.moro@mbcc-group.com

R. L. Zerbino
LEMIT-CIC and Faculty of Engineering UNLP, La Plata, Argentina
e-mail: zerbino@ing.unlp.edu.ar

R. Gettu
Department of Civil Engineering, Indian Institute of Technology Madras, Chennai, India
e-mail: gettu@iitm.ac.in

H. Pauwels
NV BEKAERT SA, Zwevegem, Belgium
e-mail: hans.pauwels@bekaert.com

T. Nishiwaki
Department of Architecture and Building Science, School of Engineering, Tohoku University,
Sendai, Japan
e-mail: tomoya.nishiwaki.e8@tohoku.ac.jp

B. Parmentier
BBRI Belgium Building Research Institute, Limelette, Belgium
e-mail: benoit.parmentier@bbri.be

A. Llano-Torre and P. Serna (eds.), *Round-Robin Test on Creep Behaviour
in Cracked Sections of FRC: Experimental Program, Results and Database Analysis*,
RILEM State-of-the-Art Reports 34, https://doi.org/10.1007/978-3-030-72736-9_5

was directly provided by the participants and follows the same structure for each participant where institution identification and contact information was included. Each participant summarised the most significant data such as specimen's size, load configuration, parameters or environmental conditions in a quite useful table. Moreover, the participants included description and close-up pictures of the creep frames construction, the support boundary conditions, and the measurement devices used. Finally, the participants provided a complete description of the specific complete creep test procedure followed.

5.1 Description of Equipment/Procedures: LAB-01

Institution/company:	Universitat Politècnica de València
Department:	ICITECH Instituto de Ciencia y Tecnología del Hormigón
Information provided by:	Aitor Llano-Torre
E-mail address:	aillator@upv.es

N. Buratti
Department of Civil, Chemical, Environmental and Materials Engineering DICAM, University of Bologna, Bologna, Italy
e-mail: nicola.buratti@unibo.it

R. D. Toledo Filho
Civil Engineering—COPPE, Federal University of Rio de Janeiro, Rio de Janeiro, Brazil
e-mail: toledo@coc.ufrj.br

J.-P. Charron
Department of Civil, Geological and Mining Engineering, Polytechnique Montreal, Montreal, Canada
e-mail: jean-philippe.charron@polymtl.ca

C. Larive
Centre d'Etudes des Tunnels CETU, Bron, France
e-mail: catherine.larive@developpement-durable.gouv.fr

W. P. Boshoff
Faculty of Engineering, Built Environment and Information Technology, University of Pretoria, Pretoria, South Africa
e-mail: billy.boshoff@up.ac.za

E. S. Bernard
TSE Technologies in Structural Engineering Pty Ltd, Sydney, Australia
e-mail: s.bernard@tse.net.au

M. Kompatscher
VSH VersuchsStollen Hagerbach AG, Flums Hochwiese, Switzerland
e-mail: mkompatscher@hagerbach.ch

Test method: (Flexural/Direct Tension)	Flexural
Specimens type: (Moulded, Sawed, Cores)	Moulded
N° of creep frames available:	4
N° of specimens by frame:	3
Specimens dimensions: (Prismatic, cylindrical, square panel or round panel)	Prismatic, $150 \times 150 \times 600$ mm
Notch: (yes/no)	Yes
Creep Test configuration: (Only for flexural beam test: 3PBT/4PBT)	Four-point bending test (4PBT)
Load application rollers: (In the case of 3PBT, only roller A)	OXX—free Rotation X, fixed Rotation Z, fixed Translation X XOX—fixed Rotation X, free Rotation Z, fixed Translation X
Supporting rollers: (rollers A and B freedom description)	OXX—free Rotation X, fixed Rotation Z, fixed Translation X XOX—fixed Rotation X, free Rotation Z, fixed Translation X
Load transfer: (lever arm, hydraulic, dead load...)	Dead load in lever arm
Pre-crack level:	CMOD equal to 0.5 mm
Load level:	50% of $f_{R,p}$
Load calibration:	By load cell
Load control in time:	Weekly controlled and adjusted
Transducers:	1 LVDT per specimen and 1 Load cell per frame
Transducer position:	LVDT under the specimen
DAS rate: (Continuous, diary, weekly...)	Continuous
Displacements measured: (CMOD, COD, CTOD or deflection)	CMOD
Frames location: (climatic room, laboratory or outside)	Climatic room
Temperature controlled: (yes/no)	Yes, 20 °C $\pm$ 1 °C
Humidity controlled: (yes/no)	No, 49% $\pm$ 16%

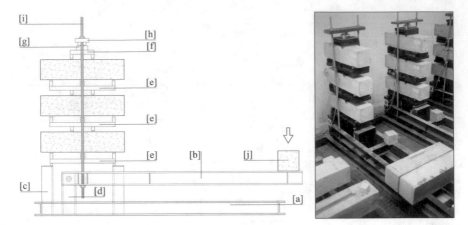

Fig. 5.1 Creep frame geometry and view of frames during creep test

5.1.1 Description of Creep Frames

The flexural creep test frames were designed to test in creep up to three prismatic specimens [30, 31] as seen in Fig. 5.1. The frames are built in steel and consist of a second-degree lever arm system where deadload counterweights (j) are placed over the lever arm (b) to induce the required stress in the cracked section. The applied load is transferred to the top specimen by means of two screwed bars (i). A load transfer steel plate (h) is located just over the load cell (g) that measures the load induced. The load cell lays on the short load transfer element (f) that generates two loading points in the specimen. The support rollers are in the large load transfer plate (e) located between specimens generating a four-point bending test (4PBT) load configuration.

The creep tests are performed in a climatic room where environmental conditions are continuously monitored by datalogger connected to the data acquisition system (DAS). The temperature is kept stable during the long-term test in 20 °C $\pm$ 2 °C while the relative humidity inside the climate chamber is only restricted and ranges around 48% $\pm$ 20% depending on the season.

5.1.2 Support Conditions

In order to allow an adequate adjustment of the column of specimens and compensate geometry errors, both the loading and the support rollers have one of the rollers fixed whereas the other roller can turn in a plane orthogonal to the longitudinal axis of the specimen as seen in Fig. 5.2.

At least one of the rollers has bearings that make possible the free rotation during the crack mouth opening displacement (CMOD) deformations, avoiding constrictions due to the boundary conditions.

Fig. 5.2 Support rollers: **a** free turn roller (XOX), **b** free rotation roller (OXX)

Fig. 5.3 a Load cell location between load steel plate and the short load transmission element, **b** LVDT transducers location

5.1.3 Measuring System and Data Acquisition

The applied load on each creep frame is continuously monitored by one load cell (see Fig. 5.3a) located over the top specimen of the frame (g). Delayed CMOD are measured by linear variable displacement transducers (LVDT) located at the bottom side of each specimen (see Fig. 5.3b) connected to the DAS. The applied load on each the frames is weekly checked. If any variation in the load are observed due to internal frictions of the system, the counterweight is adjusted to the required load for the $f_{R,c}$.

5.1.4 Creep Test Procedure

Pre-cracking tests of the specimens were performed in compliance with EN 14651 [22] standard recommendations. The specimens were notched and tested in a three-point bending test (3PBT) configuration in 500 mm length support span at 0.5 mm/min CMOD rate. Once the target nominal $CMOD_{pn}$ of 0.5 mm was reached, the specimens were unloaded at 2 mm/min rate. Once the load was totally removed, both elastic and delayed recovery deformations 10 min after unloading were recorded. Then, the specimens were removed and stored supported on their side (turned 90°) waiting to be placed in the creep frames. Although the multi-specimen setup allows to stack up to three specimens, five specimens of each batch were pre-cracked so that the best subset of specimens with minimum creep index variation could be selected.

The specimens were stored two days before the beginning of the creep test in the climate room where the creep test would be performed. Thereby, the specimens had time enough to acclimate to the test environmental conditions of the climate room to minimise interfering with the instantaneous and short-term readings.

Due to the multi-specimen configuration of the frames, a 4PBT load configuration was used during long-term tests to improve the stability of the specimens. This change from 3PBT in the pre-cracking test to the 4PBT in the creep test required a new formulation to obtain the corresponding load for each frame. The relative position (top, middle or bottom) of each specimen in the column was decided considering their mechanical performance and the deadload of the upper specimens and other transfer elements. Hence, specimens with higher residual strengths were located at the bottom of the column to achieve similar *creep index* for all the specimens of each frame.

The mounting of the specimens was gradually done. Once the first specimen was placed in the frame, the LVDT transducer was connected to the DAS and the first zero registered. After that, the first load transmission element was installed together with the second specimen. Since the LVDT of the bottom specimen was already connected, the CMOD due to the weight of the load transmission element and the next specimens was recorded. Once the second LVDT was connected and the second zero recorded, the next load transfer element for the last specimen was installed, the third specimen placed and the LVDT was connected to measure the CMOD. The last load transmission element was placed over the last specimen together with the load cell and the load plate. This load transfer plate was connected to the lever arm by means of two screwed bars. The lever arm was lifted by means of a manual hydraulic jack and then the load plate connected to the screwed bars by nuts. These nuts were equally tightened up to 1 kN of load measured by the load cell and then the hydraulic jack was gradually removed. The first step of load (usually 2–3 kN) was induced by the dead weight of the lever arm. Then, the counterweights were located at the end of the lever until the desired load was reached in the load cell. The load could be adjusted either by adding more weight or moving the counterweight trough the lever arm. The load was continuously monitored since load reduction in time may occur due to any internal friction of the steel frame components. The load of each frame was weekly checked and adjusted if required to assure the desired load in the frames during the 360 days of the creep test.

After 360 days of sustained load, creep frames were unloaded but the specimens were kept in place stacked for 30 additional days to register delayed recovery deformations. The lever arm was gradually lifted by using a hydraulic jack, the nuts of the screwed bars were untightened, and the counterweights were removed from the lever. After the 30 days delayed recovery period, the LVDT of the top specimen was disconnected and the specimen and the load transfer element removed. After that, the second LVDT was disconnected and the second specimen removed and stored. Finally, the last specimen is disconnected and removed. Note that all the specimens were stored supported on their side turned 90° to avoid any CMOD variation due to the dead load of the specimens before the last post-creep flexure test.

The post-creep flexure test was performed in compliance with EN 14651. A 3PBT load configuration was then performed and, since the specimens were already cracked, the applied loading rate was 2 mm/min of CMOD. Following the agreed procedure for the RRT, one hysteresis loop was performed at the sustained stress during the creep test $f_{R,c}$ and then, the test continued up to 4 mm CMOD was achieved. After the test, the specimens were broken into two slices to proceed with the counting of the fibres crossing the cracked section.

5.2 Description of Equipment/Procedures: LAB-02

Institution/company:	Universitat Politècnica de Catalunya (UPC—Barcelonatech)
Department:	Civil and Environmental Engineering
Information provided by:	Sergio Cavalaro
E-mail address:	S.Cavalaro@lboro.ac.uk

Test Method: (Flexural/Direct Tension)	Flexural
Specimens type: (Moulded, Sawed, Cores)	Moulded
N° of creep frames available:	12
N° of specimens by frame:	3
Specimens dimensions: (Prismatic, cylindrical, square panel or round panel)	Prismatic (150 mm × 150 mm × 600 mm)
Notch: (yes/no)	Yes
Creep Test configuration: (Only for flexural beam test: 3PBT/4PBT)	Four-point bending test (4PBT)
Load application rollers: (In the case of 3PBT, only roller A)	XXX—fixed Rotation X, Rotation Z and Translation X XXX—fixed Rotation X, Rotation Z and Translation X
Supporting rollers: (rollers A and B freedom description)	XXX—fixed Rotation X, Rotation Z and Translation X XXX—fixed Rotation X, Rotation Z and Translation X

(continued)

(continued)

Load transfer: (lever arm, hydraulic, dead load...)	Lever arm
Pre-crack level:	CMOD equal to 0.5 mm
Load level:	50% of $f_{R,p}$
Load calibration:	Dead load applied to lever arm
Load control in time:	Beginning of the test and every week
Transducers:	Digital DEMEC mechanical strain gauge with metallic reference points glued to the surface
Transducer position:	2 mm from crack mouth
DAS rate: (Continuous, diary, weekly...)	2 times a day for first week, then every week
Displacements measured: (CMOD, COD, CTOD or deflection)	CMOD
Frames location: (climatic room, laboratory or outside)	Laboratory
Temperature controlled: (yes/no)	No
Humidity controlled: (yes/no)	No

5.2.1 Description of Creep Frames

Figure 5.4 shows a side scheme of the Creep Frame and a the loaded creep frames in the laboratory, including its main components and the specimens defined in accordance with the setup proposed by Arango et al. [6]. The loading mechanism was based on the lever principle: applying a force in one edge of a rigid object used with an appropriate fulcrum or pivot point can multiply the mechanical force applied to

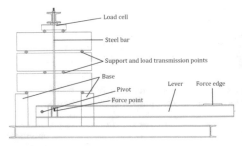

Fig. 5.4 Creep frame geometry and view of frames during creep test

Fig. 5.5 Specimen during the creep test (**a**) and detail of the rollers used as support and to apply a load (**b**)

another object. The base of the frame and the fulcrum are made of steel I profiles. The lever is connected to the base of the frame at the pivot point that allows the lever to rotate. On the opposite end, the lever has a force edge where a known weight is placed to load the setup. Closer to the pivot point, 2 steel bars connect the lever (one at each side) and one I profile located above the specimens. Specimens are staked one over the other in a pile of 3. The weight placed in the force edge of the lever pushes the steel bar downwards and apply a load to the specimens. The top specimen receives the load from the top I profile. Several rollers placed between the specimens transport the load to the base of the frame. A load cell is placed between the top I profile and the top specimen to register and control the load applied during the test.

5.2.2   Support Conditions

Figure 5.5 shows de detail of the steel rollers that act as support and load application elements in a 4PBT setup adopted for the creep test. The rollers consist of steel cylinders placed between the specimens. Since both specimens turn in the same direction, the spin of the rollers is compensated. Therefore, the spin and the translation are fixed in both the longitudinal axis and in the transversal direction.

5.2.3   Measuring System and Data Acquisition

The CMOD was measured with a Digital DEMEC mechanical strain gauge with a length of 100 mm and accuracy of 8 micro-strains. Stainless-steel reference points were fixed using a 2-compound fast-setting epoxy resin. The 2 points were placed following the main axis of the beam, crossing the notch and approximately 2 mm apart from the crack mouth. The Digital DEMEC mechanical strain gauge was calibrated before each measurement (see Fig. 5.6a). The load cell was placed above the top beam, as shown in Fig. 5.6b. The load was controlled daily during the first week and once every two weeks after that.

Fig. 5.6 Digital DEMEC mechanical strain gauge (**a**) and load cell above the top beam (**b**)

5.2.4 Creep Test Procedure

First of all, the beams were notched to a depth of 25 mm, the stainless-steel points for measurement of the crack opening were fixed, and pre-cracking was performed following the procedure specified in EN 14651. The pre-cracking was performed in a 3PBT configuration with 500 mm distance between supports and load applied to midspan. One support was free to turn in the longitudinal direction, and the other was fixed. All supports and load application elements were free to rotate in the transversal direction. The CMOD was controlled during the pre-cracking, and the load applied to each beam was registered. The loading procedure consisted of a first ramp with CMOD rate of 0.5 mm per minute until the CMOD of 0.5 mm was reached. After that, the specimens were unloaded at 3 mm/min. Once unloaded, specimens were turned 90° and kept in laboratory conditions until the creep test. All specimens were pre-cracked in 2 consecutive days and stored for 14 days before the creep test. Specimens were classified in ascending order of the load registered for CMOD of 0.5 mm and grouped in sets of 3 that presented the closest load. This procedure intends to guarantee that the load level of all specimens is as close as possible to 50% of $f_{R,p}$.

The creep test procedure was divided into two stages: the loaded stage and the unloaded stage. Before the loading starts, the lever of the steel frame was lifted. Then, the support elements were placed and above them the first beam. Afterwards, the second element of support and load transmission was placed on the top surface of the first beam and above it the second beam. This procedure was repeated until the column of three beams was assembled. Then, the bars on both sides of the column were attached but not fixed (no load was transmitted to the beams at this point). The initial crack opening was measured for all beams, the bars on both sides of the column were fixed, the lever of the steel frame was liberated and, finally, the weight specified for the test was then applied. The loading stage of each creep frame took approximately 5 min. The crack opening of the beam was measured again after loading stage was completed. The load and the crack opening were controlled daily during the first week and every week after that. Corrections to the load were introduced to compensate to variations induced by the crack opening during the test.

Once the creep test was finished, specimens were unloaded and left to rest for 45 days. Specimens were, then, characterized up to failure in a 3PBT test configuration following the EN 14651 with the same setup adopted in the pre-cracking stage. A CMOD rate of 2 mm per minute was adopted in this case as the beams are already cracked. The specimens were stored in laboratory conditions during the whole experimental program (temperature and humidity were not controlled).

5.3 Description of Equipment/Procedures: LAB-03

Institution/company:	OTH REGENSBURG
Department:	Civil engineering
Information provided by:	Tobias Preischl
E-mail address:	tobias.preischl@oth-regensburg.de

Test method: (Flexural/Direct Tension)	Flexural
Specimens type: (Moulded, Sawed, Cores)	Moulded
Nº of creep frames available:	12
Nº of specimens by frame:	1
Specimens dimensions: (Prismatic, cylindrical, square panel or round panel)	Prismatic $150 \times 150 \times 500$
Notch: (yes/no)	No
Creep Test configuration: (Only for flexural beam test: 3PBT/4PBT)	Four-point bending test (4PBT)
Load application rollers: (In the case of 3PBT, only roller A)	XXX—fixed Rotation X, Rotation Z and Translation X XXX—fixed Rotation X, Rotation Z and Translation X
Supporting rollers: (rollers A and B freedom description)	XOO—fixed Rotation X, free Rotation Z, free Translation X XXX—fixed Rotation X, Rotation Z and Translation X
Load transfer: (lever arm, hydraulic, dead load...)	Dead load in lever arm
Pre-crack level:	Deflection of 0.5 mm, span, 450 mm
Load level:	50% of load at 0.5 mm deflection (OEBV)

(continued)

(continued)

Load calibration:	By balance
Load control in time:	Dead weight cannot change, no further control
Transducers:	Dial gauges (dial indicator)
Transducer position:	Both sides of beams attached to Japanese Yoke
DAS rate: (Continuous, diary, weekly…)	Flexible, starting every minute, later on every week
Displacements measured: (CMOD, COD, CTOD or deflection)	Deflection
Frames location: (climatic room, laboratory or outside)	Laboratory basement
Temperature controlled: (yes/no)	No
Humidity controlled: (yes/no)	No. Beams were protected from drying out by aluminium sheets

5.3.1 Description of Creep Frames

The long-term test was carried out on six special test rigs. The creep rigs are designed similarly to the testing machine used for the flexural test, but the load is kept at a constant level by simple leverage (Fig. 5.7). They are built in steel and consist on a second degree lever arm system where a dead load or counterweight is located in the lever arm to produce the needed load in the frame [32, 33, 34]. This load is transmitted to the top of the specimens by means of two screwed bars where a load construction is located. The creep test rigs are situated in the basement of the lab, not in climate chamber as seen in Fig. 5.8.

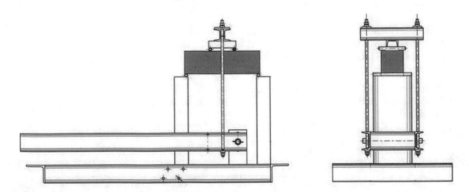

Fig. 5.7 OTH Regensburg University creep rig geometry

Fig. 5.8 Creep test rigs from LAB-03: wrapped specimen (**a**), fulcrum (**b**) and location in the basement of the laboratory (**c**)

5.3.2 Support Conditions

In order to allow a good adjustment of specimens and be able to compensate geometry errors, one of the support rollers is fixed whereas the other roller is able to turn in the longitudinal axis of the specimen (Fig. 5.9). Both loading rollers are welded as seen in Fig. 5.10c.

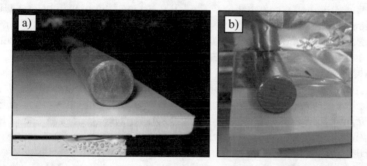

Fig. 5.9 View of one of the support rollers (XOO) of different creep rigs

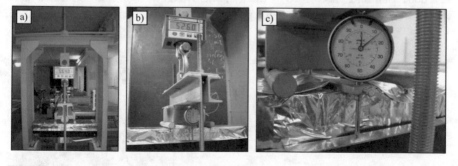

Fig. 5.10 Load calibration and dial gauges location

5.3.3 Measuring System and Data Acquisition

The dead load is adjusted by a balance whereas the delayed deformation readings come from two dial gauges at midspan deflection of each beam (Fig. 5.10).

5.3.4 Creep Test Procedure

Testing for this investigation was performed on beams with a span of 450 mm, applying 4-point flexural loading in accordance with the testing procedure from the Guideline for Fibre Reinforced Concrete published by the Austrian Society for Concrete and Construction Technology (OEBV) [24] as illustrated in Fig. 5.11. For the purpose of this investigation, unnotched $150 \times 150 \times 600$ mm beams were first tested under 4-point flexural loading up to a deflection of 0.5 mm. Loading rate 0.02 mm/min up to 0.5 mm, then 0.2 mm/min. The beams were then unloaded and three pre-cracked beams were loaded with a sustained load at 50% of the load reported at 0.5 mm deflection. Note that the procedure for creep testing according to OEBV was modified somewhat to fit the RRT: Preload unnotched beams at 0.5 mm deflection, 0.5 mm for the 500 mm span of EN 14651 should be here 0.45 mm.

Six creep test rigs were used for these tests. They are situated in the basement of the lab, not in a climate chamber. Aluminium sheets were used to protect the beams from drying out.

Each beam is being loaded separately, as the loads of the beams differ. The creep deformation versus time was registered by dial gauges. Unloading took place after one year and recovery was registered for 39 days. The specimens tested in creep were then remove and tested to failure.

The post-creep failure test of the specimens was performed following Austrian Guideline Fibre reinforced concrete standard recommendations [24]. For these tests, a 4PBT load configuration is again adopted and the load rate is now 0.15 mm/min mid deflection rate. As decided by the TC, one hysteresis loop was performed until the specific load achieved during creep test and then, the failure test continued until

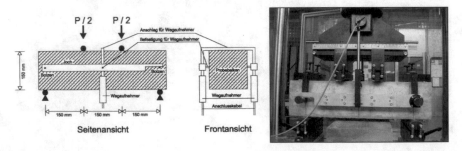

Fig. 5.11 Beam geometry, loading and deflection measurement during precracking by OEBV

Fig. 5.12 Broken beam for fibre counting after the creep test

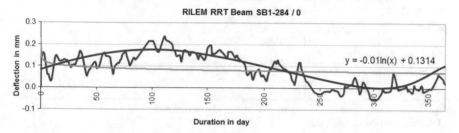

Fig. 5.13 Example of fitting curve of creep test raw data

failure. As our testing machine was moved at that time, we could not follow fully the required test regime. Therefore, we loaded once more up to the creep load. But it was not possible to register the following unloading.

When the mid-deflection reached 4 mm, the test is finished and the specimen is totally broken in order to have both sides of the cracked section ready to count the fibres. The number of fibres in the cross section is essential to interpret results. Fibre counting was done manually using a given pattern. Each face of the crack was divided into three areas with the same dimensions (Fig. 5.12). Some fibres were pulled out and some failed. The failed fibres were counted twice due to the fact that one part of the failed fibre is in every face of the crack. There was not any heterogeneous distribution visible.

For calculating the deflections over time, the final trend-line was used (Fig. 5.13). Due to the low stress-level the main part of the deflection is due to temperature. Red Trend-line would be more appropriate. The red line reproduces the small change in temperature during the year in our basement more intense than the very small creep deformation.

5.4 Description of Equipment/Procedures: LAB-04

Institution/company:	BASF CONSTRUCTION CHEMICALS ITALIA SPA
Department:	Technological Laboratory
Information provided by:	Sandro Moro
E-mail address:	sandro.moro@mbcc-group.com

Test method: (Flexural/Direct Tension)	Flexural
Specimens type: (Moulded, Sawed, Cores)	Moulded
Nº of creep frames available:	4
Nº of specimens by frame:	3
Specimens dimensions: (Prismatic, cylindrical, square panel or round panel)	Prismatic 150 × 150 × 600 mm
Notch: (yes/no)	Yes
Creep Test configuration: (Only for flexural beam test: 3PBT/4PBT)	Four-point bending test (4PBT) 150 | 150 | 150 450 150
Load application rollers: (In the case of 3PBT, only roller A)	XOX—fixed Rotation X, free Rotation Z, fixed Translation X OXX—free Rotation X, fixed Rotation Z, fixed Translation X
Supporting rollers: (rollers A and B freedom description)	OXX—free Rotation X, fixed Rotation Z, fixed Translation X XOX—fixed Rotation X, free Rotation Z, fixed Translation X
Load transfer: (lever arm, hydraulic, dead load…)	Dead load in lever arm
Pre-crack level:	CMOD = 0.5 mm
Load level:	50% of $f_{R,p}$
Load calibration:	By Load cell
Load control in time:	Weekly controlled and adjusted
Transducers:	LVDT
Transducer position:	Under the specimen

(continued)

(continued)

DAS rate: (Continuous, diary, weekly…)	Continuous
Displacements measured: (CMOD, COD, CTOD or deflection)	CMOD
Frames location: (climatic room, laboratory or outside)	Climatic room
Temperature controlled: (yes/no)	Yes, 20° C $\pm$ 2 °C
Humidity controlled: (yes/no)	Yes, 50% $\pm$ 15%

5.4.1 Description of Creep Frames

The frames were designed to test in creep up to three prismatic specimens as seen in Fig. 5.14. They are built in steel and consist of a second-degree lever arm system where a dead load or counterweight is located in the lever arm to produce the needed load in the frame. This load is transmitted to the top of the specimens by means of two screwed bars where a load steel plate is located just over the load cell which checks the load induced. The load cell lays on the short load transmission element which generates two loading points in the specimen. The specimens are supported by the large load transmission elements which help to the stabilization of the column generating a 4PBT load configuration between specimens.

Fig. 5.14 Creep frame geometry

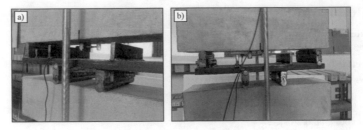

Fig. 5.15 Transfer plates with supports and emergency brakes

The creep frames and the DAS are located in a climatic room where temperature is controlled and humidity restricted. The temperature during the year is 20 °C ± 2 °C and the relative humidity inside the climate chamber is usually around 50% and varies ±15%. In both cases, the temperature and relative humidity inside the chamber were continuously registered by a separate data logger.

5.4.2 Support Conditions

Both for the load and for the support rollers of the probe, one of the rollers is fixed whereas the other roller must be able to turn in a plane orthogonal to the longitudinal axis of the specimen as seen in Fig. 5.15.

5.4.3 Measuring System and Data Acquisition

The load in each frame is controlled by one load cell located in the upper area of the frame. In order to measure the delayed CMOD deformations in each specimen, LVDTs are located at the bottom of each specimen, as seen in Fig. 5.16, and connected to the DAS. The load in the frames is weekly checked and if slight variations in the load are observed due to small internal frictions of the system, the counterweight is readjusted to adjust the load to the desired one.

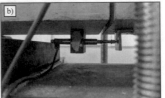

Fig. 5.16 Load cell (**a**) and the LVDT (**b**)

5.4.4 Creep Test Procedure

The pre-cracking test of the specimens was performed following the EN 14651 standard recommendations. The specimens where notched and tested in a 3PBT configuration in 500 mm length span at 0.5 mm/min CMOD rate. When the desired CMOD of 0.5 mm was achieved, the specimens were unloaded at 0.5 mm/min. Once the specimen was totally unloaded, the recovery deformations were recorded for 10 min and then the specimen was removed and stored turned 90° waiting to be located in the creep frames.

Previously to the beginning of the creep test, the specimens were stored for two days in the climate room where the creep test will be performed. Thereby, the specimens have at least one day to acclimate to the environmental conditions of the climate room and we avoid any interference with the instantaneous readings of the thermal acclimation.

Due to the multiple specimen setup of our creep frame, the creep tests are performed in a 4PBT configuration to improve the stabilization of the column. Therefore, a numerical transformation must be realized to obtain the load of each frame in a 4PBT load configuration. The relative position of each specimen in the column is decided keeping in mind the load of the upper specimens and the desired load level. Thereby, at the bottom of the column are usually placed specimens with higher residual strengths and thus, it is possible to achieve very similar *creep index* for the specimens of each frame.

The mounting of the specimens must be done step by step. All three specimens were mounted on the frame. Above the last specimen must be placed the last load transmission element and then the load cell and the load plate. This load plate is connected to the lever arm by means of two screw bars. Then the first zero was registered.

The dead weight of the lever arm makes the first step of load and then dead load is located as counterweight until the desired load is reached in the load cell. The load is adjusted by adding more weight or moving the counterweight trough the lever arm. Due to internal frictions of the steel frame, the load can be reduced along time. Thus, the load cells readings were weekly checked, and the load adjusted if needed in order to assure the required load in the frames during the 360 days duration of the creep test.

Once the creep test is finished, the frames are unloaded and the specimens left in place for 30 days in order to register recovery deformations after creep test. The lever arm is gradually lifted by means of a hydraulic jack and the nuts of the screw bars are untightened. Thus, the external load of the frame is removed and only remains the weight load of the upper specimens. After one month recording recovery deformations, the demounting process starts in a similar way than when mounting. The upper specimens LVDT is disconnected and the specimen removed and store turned 90° waiting to be tested in flexure. When the load transmission element is removed and there is no weight over the second specimens, the second LVDT can be disconnected and the second specimen removed and stored. In a similar way, the last specimen is disconnected and removed when all weight elements were removed. All the specimens are stored turned 90° in order to avoid CMOD displacement due to the dead load of the specimens waiting to be tested again in flexure.

The post-creep failure test of the specimens tested in creep was performed following the EN 14651 standard recommendations. For these tests, a 3PBT load configuration is again adopted and the load rate is now 2 mm/min CMOD rate since the specimens are already cracked. As decided by the TC, one *hysteresis loop* was performed until the specific load achieved during creep test and then, the failure test continued until failure. When the CMOD achieves 4 mm, the test is finished, and the specimen is totally broken in order to have both sides of the cracked section ready to count the fibres.

5.5   Description of Equipment/Procedures: LAB-05

Institution/company:	LEMIT-CIC and Facultad de Ingeniería UNLP
Department:	Civil Engineering
Information provided by:	Raúl Zerbino
E-mail address:	zerbino@ing.unlp.edu.ar

Test method: (Flexural/Direct Tension)	Flexural
Specimens type: (Moulded, Sawed, Cores)	Moulded
N° of creep frames available:	4
N° of specimens by frame:	3
Specimens dimensions: (Prismatic, cylindrical, square panel or round panel)	Prismatic, 150 × 150 × 600 mm
Notch: (yes/no)	Yes
Creep Test configuration: (Only for flexural beam test: 3PBT/4PBT)	Four-point bending test (4PBT)
Load application rollers: (In the case of 3PBT, only roller A)	OXX—free Rotation X, fixed Rotation Z, fixed Translation X XXX—fixed Rotation X, Rotation Z and Translation X

(continued)

(continued)

Supporting rollers: (rollers A and B freedom description)	OXX—free Rotation X, fixed Rotation Z, fixed Translation X XOX—fixed Rotation X, free Rotation Z, fixed Translation X
Load transfer: (lever arm, hydraulic, dead load…)	Dead load in lever arm
Pre-crack level:	CMOD equal to 0.5 mm
Load level:	50% of $f_{R,p}$
Load calibration:	By load cell
Load control in time:	No
Transducers:	1 Electronic LVDT/specimen + 1 Mechanical gage /specimen
Transducer position:	LVDT at CTOD, Mechanical gage at CMOD
DAS rate: (Continuous, diary, weekly…)	LVDT: continuous during at least 15 days Mechanical gage: diary (15 days) and 3 days a week (1 year)
Displacements measured: (CMOD, COD, CTOD or deflection)	LVDT at CTOD, Mechanical gage at CMOD
Frames location: (climatic room, laboratory or outside)	Climatic room
Temperature controlled: (yes/no)	Yes, 22 °C ± 3 °C
Humidity controlled: (yes/no)	No. Variable between 40 and 85% along the year depending the season

5.5.1 Description of Creep Frames

The specimens were placed in creep frames [35] where the sustained loads were applied using a 4PBT configuration, a multiple specimen's setup in column was adopted for the creep period. Figure 5.17 shows the load configuration inside the creep frame. The specimens with highest residual strength were placed at the bottom of the column to minimize differences in stress levels due to weight of the upper specimens, in this way, the applied stresses were similar.

5.5.2 Support Conditions

Both for the loading and for the support rollers of the probe, one of the rollers is fixed whereas the other roller must be able to turn in a plane orthogonal to the longitudinal axis of the specimen (Fig. 5.18).

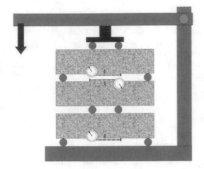

Fig. 5.17 Creep frame geometry and load configuration

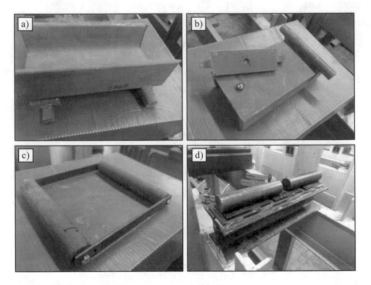

Fig. 5.18 Load and support rollers boundary conditions: **a** and **b** upper loading supports, **c** intermediate loading supports and **d** bottom supports

5.5.3 Measuring System and Data Acquisition

The load in each frame was controlled by one load cell located in the upper area of the frame during the first loading, with simultaneous measurements of CTOD (LVDT) and CMOD (dial gauge), up to obtaining the desired value by incorporating an appropriated dead load. Then the load was released, the load cell was replaced by a steel block with the same dimensions, and the load was applied again. During applying loads the CTOD was continuously measured in each specimen by one LVDT connected to a DAS which continuously records the load cell and the transducers readings; LVDTs measurements were continued during at least 15 days. During the creep test duration, the CMOD evolution was registered by means of one dial gauge. Transducers location can be observed in Fig. 5.19.

Fig. 5.19 Dial gauges transducer location: **a** dial gauge for CMOD and **b** LVDT for CTOD

5.5.4 Creep Test Procedure

The pre-cracking test of the specimens was performed following the EN 14651 standard recommendations. The specimens where notched and tested in a 3PBT configuration in 500 mm length span at 0.5 mm/min CMOD rate. The test was controlled by measuring the CMOD through a clip-gage. When the desired CMOD of 0.5 mm was achieved, the specimens were unloaded. The recovery deformations were recorded for 5 min and then the specimen was removed and stored turned 90° waiting to be located in the creep frames.

Previously to the beginning of the creep test, the specimens were stored for two days in the climate room where the creep test will be performed. Thereby, the specimens have at least one day to acclimate to the environmental conditions of the climate room and we avoid any interference with the instantaneous readings of the thermal acclimation.

After 1 year under sustained loads (creep period), the specimens were unloaded and kept in the frames while the CMOD recovery was measured for 30 days. Afterward, the specimens were removed from the creep frames (again the specimens are stored turned 90° in order to avoid CMOD displacement due to their dead load) and then tested until failure using a 3PBT configuration following the general guidelines of EN 14651 standard. As decided by the TC, one hysteresis loop was performed until the specific load achieved during creep test and then, the failure test continued until failure. Finally, the number of fibres at the fracture surface was counted and the density of fibres was calculated.

5.6 Description of Equipment/Procedures: LAB-06

Institution/company:	Indian Institute of Technology Madras IITM
Department:	Civil Engineering
Information provided by:	Ravindra Gettu
E-mail address:	gettu@iitm.ac.in

Test method: (Flexural/Direct Tension)	Flexural
Specimens type: (Moulded, Sawed, Cores)	Moulded
N° of creep frames available:	4
N° of specimens by frame:	3 specimens in a frame
Specimens dimensions: (Prismatic, cylindrical, square panel or round panel)	Prismatic, $150 \times 150 \times 600$ mm
Notch: (yes/no)	Yes (25 mm deep, 3 mm wide, at mid-span)
Creep Test configuration: (Only for flexural beam test: 3PBT/4PBT)	Four-point bending test (4PBT) *175* *150* *150* *500* *150*
Load application rollers: (In the case of 3PBT, only roller A)	OOX—free Rotation X, free Rotation Z, fixed Translation X OOX—free Rotation X, free Rotation Z, fixed Translation X
Supporting rollers: (rollers A and B freedom description)	OOX—free Rotation X, free Rotation Z, fixed Translation X OOX—free Rotation X, free Rotation Z, fixed Translation X
Load transfer: (lever arm, hydraulic, dead load…)	Lever arm
Pre-crack level:	0.5 mm
Load level:	48 to 65%
Load calibration:	Yes
Load control in time:	Yes
Transducers:	Clip gauge
Transducer position:	Clip gauge mounted across the notch
DAS rate: (Continuous, diary, weekly…)	Continuous
Displacements measured: (CMOD, COD, CTOD or deflection)	CMOD
Frames location: (climatic room, laboratory or outside)	Climatic room
Temperature controlled: (yes/no)	Yes (25 °C ± 2 °C)
Humidity controlled: (yes/no)	Yes (65% $\pm$ 5%)

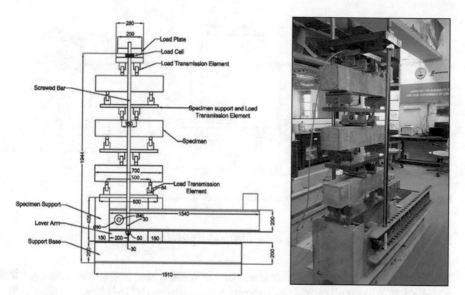

Fig. 5.20 Schematic illustration and view of flexural creep test frame

5.6.1 Description of Creep Frames

To study the flexural creep of cracked FRC, a second-order lever system was used with a load multiplying factor of about 15. The details of the flexural creep frame and the view of the frame with specimens are given in Fig. 5.20. The set of test specimens is supported on two vertical profiles (I-sections) bolted to the longitudinal sections on the support base. The base and lever system (placed on either side of the specimens) are made up of steel channel sections, which are longitudinal, with stiffeners placed transversely to maintain the channels parallel. The load is transmitted through two threaded rods, and the lever profile has holes for allowing threaded rods to through freely. The load plate and load transmission elements are made of EN-24 steel, having yield strength of 650 MPa. The load cell is placed between the load plate and the top-most load transmission element. The design has been done to ensure negligible loss of load due to friction between the different elements of the frame.

5.6.2 Support Conditions

The support elements are placed under every specimen to provide a span of 500 mm and the load transmission elements with loading points at 150 mm are provided above each specimen. These elements are connected between two adjacent specimens. All the elements are made of rigid EN 24 steel plates. The supports and load transmission elements are designed such for free rotation and to avoid the application of torque on the specimens. The details of the base supports and rollers are shown in Fig. 5.21.

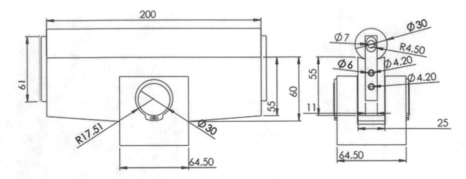

Fig. 5.21 Design of support roller load transmission element

5.6.3 Measuring System and Data Acquisition

The creep test requires the measurement of the applied load and crack opening, for which a 50 kN load cell (HBM) and a 5 mm range clip gauge (TML) are required; see Fig. 5.22. A data acquisition system (MGC plus, HBM) configured with the HBM Catman Easy software has been used, with a data sampling rate of 0.02 Hz, and the data is saved every hour.

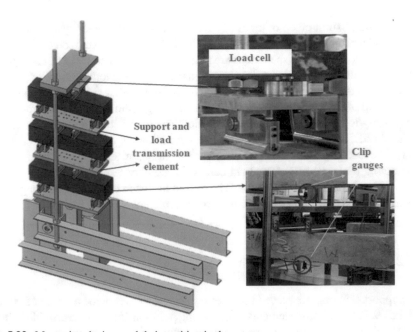

Fig. 5.22 Measuring devices and their position in the set-up

5.6.4 Creep Test Procedure

The flexural creep test was performed by applying a constant bending load on a 150 × 150 × 600 mm pre-cracked notched specimen and by monitoring the load-crack opening response for the testing period of about 1 year, followed by performing a post creep load-crack opening response. The process flow of the entire creep testing procedure is given in Fig. 5.23.

Prior to pre-cracking the specimens, a notch was cut across the width of specimen at mid-span on the face that is perpendicular to the casting direction, using a water-cooled diamond edged disc saw. The notch depth adopted for the 150 mm deep specimen was 25 mm. The specimens were maintained in the creep test environment from reception onwards. The pre-cracking tests were done two weeks after the specimens were received. The pre-cracking was done in a 300 kN capacity closed-loop servo-hydraulic testing system, wherein test is performed initially by increasing the load at a constant rate of 100 N/s up to about 40% of the estimated peak load, and then by changing to CMOD control with a rate of 2 mm/min. The load configuration for both pre-cracking and post-creep test can be observed in Fig. 5.24.

The creep load is applied in a 4PBT configuration, with about 50–60% of the nominal stress corresponding to $F_{R,p}$. The data acquisition is started and the mounting of the specimens in the creep frame is done in the following sequence: the supporting element is set on the vertical I section profile, followed by placing the first specimen over the support; subsequently, the corresponding support and load transmission element are placed on the specimen, over which the second specimen is placed; the

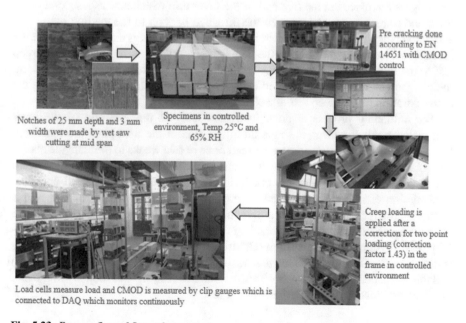

Notches of 25 mm depth and 3 mm width were made by wet saw cutting at mid span

Specimens in controlled environment, Temp 25°C and 65% RH

Pre cracking done according to EN 14651 with CMOD control

Creep loading is applied after a correction for two point loading (correction factor 1.43) in the frame in controlled environment

Load cells measure load and CMOD is measured by clip gauges which is connected to DAQ which monitors continuously

Fig. 5.23 Process flow of flexural creep test

Fig. 5.24 Setup of the
pre-cracking and post-creep
bending tests

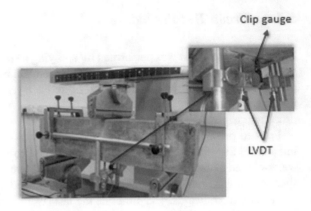

same is repeated for the third specimen. It is checked whether the specimens and the support and load transmission elements are parallel. Over the centre of the load transmission element of the third (top) specimen, the 50 kN load cell is placed. The load plate is placed on the top head of the load cell and measurement devices such as clip gauges are placed in their right positions. A temporary support is given to the free end of the lever arm which is at 0.30 m from the ground, and nuts are tightened on the threaded bars initially at the bottom and subsequently the nuts on the top of the load plate. The temporary supports are removed, and the nuts are tightened until the lever arm is horizontal, which is checked with bubble level, thereby a preload of 5 kN is applied to the free end of the lever arm.

Weights are placed in the free end of the lever arm which acts as a second order lever and transfers the load through the threaded bars on to the loading plate. The actual load applied can be controlled by the load cell readings and it depends on the *creep index* which is the ratio of the actual load applied to the specimen in creep test to the load required to pre-crack the specimen. The creep set-up is loaded and the clip gauges which are connected to the DAS give uninterrupted data during the entire duration of the creep test of 360 days.

The unloading process is done instantaneously at the free end of the lever arm which holds the lever, the nuts on either side of the load plate are loosened. Unloading–reloading cycles are performed once in two weeks to evaluate the elastic compliance evolution of the beams. In fracture mechanics, the compliance of a beam is known to be a function of the crack depth. The load plate, load cell and load transmission elements are removed, whereas the DAS is activated for two weeks to capture the recovery of the crack openings. The test specimens are then removed from the creep frame and the final bending test is done in the 300 kN capacity closed-loop servo-hydraulic testing system (Controls) (same set-up where the pre-cracking is done), in CMOD controlled manner with a rate of 2 mm/min until failure (4 mm) and the test is performed as per EN 14651 standard. A hysteresis loop is provided at the sustained load level during the bending test.

5.7 Description of Equipment/Procedures: LAB-07

Institution/company:	NV BEKAERT SA
Department:	Building Products
Information provided by:	Hans Pauwels
E-mail address:	Hans.pauwels@bekaert.com

Test method: (Flexural/Direct Tension)	Flexural
Specimens type: (Moulded, Sawed, Cores)	Moulded
N° of creep frames available:	6
N° of specimens by frame:	1
Specimens dimensions: (Prismatic, cylindrical, square panel or round panel)	Prismatic—150 × 150 × 600 mm
Notch: (yes/no)	Yes
Creep Test configuration: (Only for flexural beam test: 3PBT/4PBT)	Three-point bending test (3PBT)
Load application rollers: (In the case of 3PBT, only roller A)	XXX—fixed Rotation X, Rotation Z and Translation X
Supporting rollers: (rollers A and B freedom description)	XOO—fixed Rotation X, free Rotation Z, free Translation X XOO—fixed Rotation X, free Rotation Z, free Translation X
Load transfer: (lever arm, hydraulic, dead load…)	Dead load in lever arm
Pre-crack level:	δ of 0.475 mm ≈ CMOD equal to 0.5 mm
Load level:	50% of $f_{R,p}$
Load calibration:	By load cell
Load control in time:	-
Transducers:	Analog displacement gauge
Transducer position:	Displacement measured at bottom of specimen
DAS rate: (Continuous, diary, weekly…)	Scheduled

(continued)

(continued)

Displacements measured: (CMOD, COD, CTOD or deflection)	Deflection at midspan
Frames location: (climatic room, laboratory or outside)	Laboratory
Temperature controlled: (yes/no)	Not controlled—21 °C (SD 2 °C) during trial
Humidity controlled: (yes/no)	Not controlled—49.4% (SD 10.5%) during trial

5.7.1 Description of Creep Frames

Figure 5.25 shows a creep frame used for this test program. The creep frame is a steel construction that can accommodate a single specimen. It is a second-degree lever arm system using a dead weight (b) to introduce the required load. A steel roller (c) attached to the lever arm (a) transmits the load onto the specimen (e). Two free rolling steel bars (d) support the specimen itself.

The creep frames are located in a separate room in the laboratory where the ambient temperature and relative humidity are not actively controlled but registered during the tests.

5.7.2 Support Conditions

The two supporting rollers can roll freely; the load application roller has a flattened side to allow careful positioning of the load in the centre of the specimen. Figure 5.26 shows this principle.

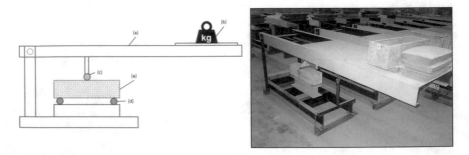

Fig. 5.25 Creep frame geometry and view of frames during creep test

Fig. 5.26 Supporting rollers

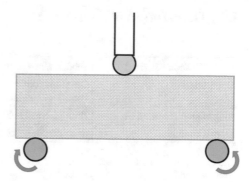

5.7.3 Measuring System and Data Acquisition

The load is initially determined using a load cell and a dummy steel element with the same size as a concrete specimen as explained below in the next section. During the actual creep test period, the load is not monitored. The deflection of each specimen is measured using analogue displacement gauges. Measurements are done according to a predefined schedule.

5.7.4 Creep Test Procedure

Specimens are notched according to EN 14651 and then pre-cracking tests are performed according to EN 14651 until 0.5 mm CMOD is reached. After unloading, the CMOD is measured.

The pre-cracked beams are installed in the creep frames. The 50% of $f_{R,p}$ load is firstly installed on the creep supports not using the test beam, but using a dummy steel beam (in order to avoid dynamic impact on the concrete test beams). A load cell is used to verify the load. This load cell is placed in between the point of load application and the dummy steel beam. The steel dummy beam is replaced by the pre-cracked concrete beam afterwards. The deflection is measured just before and just after the load application. The deflection of the beam just after the load application is the point of reference. The deflection measurements are performed manually. The sampling frequency is high in the beginning (each half hour) and decreasing afterwards (2 measurements/day). No measurements are taken in the weekend or in holidays. The environmental measurements are done automatically with a sampling frequency of 30 min.

When the creep test ends, the loads are removed in one smooth movement from the setup and the beams are kept in their position for another 46 days while the (recovery of) the deflection is continued to be measured. After that, the beams are tested according to EN 14651 until 3.5 mm deflection is reached (reference 0 = deflection at start of the EN test). The beams are first pre-loaded until the creep load (50% of $f_{R,p}$ during pre-cracking test) after which they are unloaded (until 1000 N). They are then reloaded according to the EN 14651 test.

5.8   Description of Equipment/Procedures: LAB-08

Institution/company:	Tohoku University
Department:	Department of Architecture and Building Science, Faculty of Engineering
Information provided by:	Tomoya Nishiwaki
E-mail address:	tomoya.nishiwaki.e8@tohoku.ac.jp

Test method: (Flexural/Direct Tension)	Flexural
Specimens type: (Moulded, Sawed, Cores)	Moulded
N° of creep frames available:	4
N° of specimens by frame:	2
Specimens dimensions: (Prismatic, cylindrical, square panel or round panel)	Prismatic, $150 \times 150 \times 600$ mm
Notch: (yes/no)	Yes
Creep Test configuration: (Only for flexural beam test: 3PBT/4PBT)	Four-point bending test (4PBT)
Load application rollers: (In the case of 3PBT, only roller A)	XXX—fixed Rotation X, Rotation Z and Translation X XXX—fixed Rotation X, Rotation Z and Translation X
Supporting rollers: (rollers A and B freedom description)	XOO—fixed Rotation X, free Rotation Z, free Translation X XOO—fixed Rotation X, free Rotation Z, free Translation X
Load transfer: (lever arm, hydraulic, dead load…)	Nuts on screw column were tightened to adjust bending stress
Pre-crack level:	CMOD equal to 0.5 mm
Load level:	50% of $f_{R,p}$
Load calibration:	By the stress of the screw bar columns converted from the strain gauges
Load control in time:	When the error of sustain load from the target stress exceeded 10%, the nuts on screw bars were re-tightened with the wrench and the target load was maintained
Transducers:	1 Pi-gauge/specimen 1 strain gauge/screw bar column (4 bars were used for 2 specimens)

(continued)

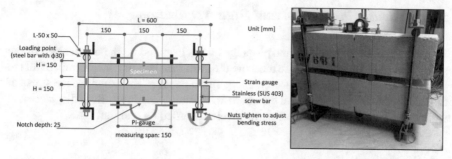

Fig. 5.27 Specimen and set-up for sustained loading

(continued)

Transducer position:	Center of the notch side of the specimen (span: 150 mm)
DAS rate: (Continuous, diary, weekly…)	Continuous
Displacements measured: (CMOD, COD, CTOD or deflection)	CMOD
Frames location: (climatic room, laboratory or outside)	Climate room
Temperature controlled: (yes/no)	Yes, 20 °C ± 3 °C
Humidity controlled: (yes/no)	Yes, 40% ± 10%

5.8.1 Description of Creep Frames

The cracked specimens were placed on a creep test set-up shown in Fig. 5.27 to apply sustained loading [36]. The sustained bending load was adjusted to 50 or 60% of the pre-cracking loading by rotating nuts at the edge of screw bar columns, and load value was calculated from the strain of the screw bar columns (SUS 304, Young's modulus: 193 kN/mm^2). During bending creep test, deformations at the tension side (bottom side) of each specimen were measured by pi-gauges. The bending creep tests were conducted in the climate room.

5.8.2 Support Conditions

The simple metal round bars were employed as the support and loading points of the specimens as shown in Fig. 5.27.

5.8.3   Measuring System and Data Acquisition

To measure the displacement, the pi-gauges were installed on the notch sides of the specimens with 150 mm span. To measure the sustained bending load, strain of all the 4 screw bar columns were measured using strain gauges. The compressive stress of the columns was converted from the strain and Young's modulus. Both displacement and bending load (strain of the column) were recorded by a digital DAS.

5.8.4   Creep Test Procedure

Pre-cracking tests were carried out in 4PBT to generate initial crack by a universal testing machine (capacity: 100kN). The deformation measured by pi-gauge was introduced up to 0.5 mm. After the bending test, cracked specimens were placed on the abovementioned creep test set-up to apply sustained loading. The sustained bending load was adjusted to 50 or 60% of the pre-cracking loading by nuts fastening. When the sustained load value become smaller than 90% of the target stress, the nuts on screw bars were re-tightened with the wrench and the target load was maintained. During bending creep test, deformations at the tension side (bottom side) of each specimen were measured by pi-gauges. The bending creep tests were conducted in the climate room. The deformation due to creep was calculated omitting the shrinkage phenomena. At the same time, other prismatic specimens with $100 \times 100 \times 400$ mm were exposed to the same condition in order to measure dry shrinkage. Regression curve of dry shrinkage was drawn according to the ACI committee 209 report as the following equation:

$$\varepsilon_{sh,t} = a \bullet \left[\frac{t}{35 + t} \right] \tag{5.1}$$

where, $\varepsilon_{sh,t}$ is dry shrinkage, t is time and a is empirical constant (246.0 for SyFRC specimens and 157.0 for SFRC specimens).

5.9   Description of Equipment/Procedures: LAB-10

Institution/company:	BBRI Belgium Building Research Institute
Department:	Structures
Information provided by:	Benoit Parmentier
E-mail address:	benoit.parmentier@bbri.be

Test method: (Flexural/Direct Tension)	Flexural
Specimens type: (Moulded, Sawed, Cores)	Moulded
N° of creep frames available:	12
N° of specimens by frame:	1
Specimens dimensions: (Prismatic, cylindrical, square panel or round panel)	Prismatic, 150 × 150 × 600 mm
Notch: (yes/no)	Yes
Creep Test configuration: (Only for flexural beam test: 3PBT/4PBT)	Three-point bending test (3PBT) 250 / 500 / 150 / 150
Load application rollers: (In the case of 3PBT, only roller A)	XOO—fixed Rotation X, free Rotation Z, free Translation X
Supporting rollers: (rollers A and B freedom description)	XOO—fixed Rotation X, free Rotation Z, free Translation X XOO—fixed Rotation X, free Rotation Z, free Translation X
Load transfer: (lever arm, hydraulic, dead load…)	Hydraulic jack
Pre-crack level:	CMOD equal to 0.5 mm
Load level:	50% of $f_{R,p}$
Load calibration:	By load cell
Load control in time:	Weekly controlled and adjusted
Transducers:	2 Electronic LVDT/specimen (1 CMOD and 1 Deflection) 1 Load cell/frame
Transducer position:	LVDT under the specimen for deflection measurement
DAS rate: (Continuous, diary, weekly…)	Continuous, higher rate during the first days
Displacements measured: (CMOD, COD, CTOD or deflection)	CMOD and Deflection
Frames location: (climatic room, laboratory or outside)	Climatic room
Temperature controlled: (yes/no)	Yes, 20 °C ± 1 °C
Humidity controlled: (yes/no)	Yes, 60% ± 5°%

Fig. 5.28 Creep loading
frames in the controlled
climatic room

5.9.1   Description of Creep Frames

The frames consist of a steel beam (IPE) supporting the concrete samples through
2 rollers as seen in Fig. 5.28. The load is maintained at a constant level by using a
piston equilibrated by a small nitrogen tank. The piston is located under the steel
beam and transmits the load in the centre of the sample through two steel rods. These
rods distribute the load over the full width of the sample through a roller. Each frame
is associated with one single sample. The frames are located in an actively controlled
room (temperature and humidity).

5.9.2   Support Conditions

The supporting rollers are two steel cylinders that can freely rotate and move placed
over the steel I profile as seen in Fig. 5.29.

Fig. 5.29 Creep frame
close-up and support rollers

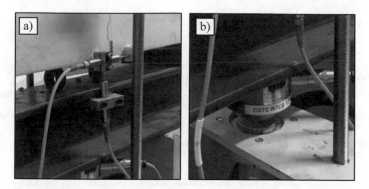

Fig. 5.30 Close-up view on the deflection and CMOD transducers (**a**) and the load cell (**b**)

5.9.3 Measuring System and Data Acquisition

The load is continuously measured by using a load cell located between the piston and the sample. The deflection and the CMOD are measured as well, with the same DAS. The data acquisition frequency follows the predefined schedule (higher rate at the start of the tests). The deflection is measured by using a LVDT at the underside of the samples, just besides the notch. The CMOD is measured by using a horizontally positioned LVDT in contact with an aluminium bracket around the notch, at the level of underside of the sample as seen in Fig. 5.30.

5.9.4 Creep Test Procedure

First of all, the samples are notched according to the procedure defined in EN 14651 up to the predefined cracking level. At the end of these tests, the distance between reference DEMEC points glued on the sides of the specimens around the notch is measured. This measurement allows for a control of the opening/closure of the crack during the displacement of the samples towards the climatic room where they will be submitted to the creep tests. After the pre-cracking test, the samples are rotated 90° around their section to avoid too much opening/closure of the crack during the displacement to the climatic room where they will be kept under constant temperature and relative humidity while submitted to a permanent load.

In the climatic room, the samples are positioned in the creep frames. The load is applied through a manual pumping system. When the load reaches the target value, the piston is connected to a small nitrogen tank that equilibrates the pressure and maintains the load constant. The level of the load is continuously monitored by the load-cell.

When the test is finished, the samples are unloaded, kept under self-weight for a short duration, rotated 90 °C and then moved to another hall to be tested according to EN 14651 until complete failure.

5.10   Description of Equipment/Procedures: LAB-11

Institution/company:	Alma Mater Studiorum, Università Di Bologna
Department:	Department of Civil, Chemical, Environmental and Materials Engineering (DICAM)
Information provided by:	Nicola Buratti
E-mail address:	nicola.buratti@unibo.it

Test method: (Flexural/Direct Tension)	Flexural Uniaxial tension
Specimens type: (Moulded, Sawed, Cores)	Moulded (flexural) Cores (uniaxial tension)
N° of creep frames available:	2
N° of specimens by frame:	3
Specimens dimensions: (Prismatic, cylindrical, square panel or round panel)	Prismatic, $150 \times 150 \times 600$ (flexural) Cylinders $\varnothing = 94$ mm $\times$ 100 mm (uniaxial tension)
Notch: (yes/no)	Yes
Creep Test configuration: (Only for flexural beam test: 3PBT/4PBT)	Four-point bending test (4PBT)
Creep Test configuration:	
Load application rollers: (In the case of 3PBT, only roller A)	OOX—free Rotation X, free Rotation Z, fixed Translation X XOX—fixed Rotation X, free Rotation Z, fixed Translation X
Supporting rollers: (rollers A and B freedom description)	OOX—free Rotation X, free Rotation Z, fixed Translation X XOX—fixed Rotation X, free Rotation Z, fixed Translation X
Supporting conditions: (for Uniaxial tension test)	Uniaxial tension with free rotation of the ends
Load transfer: (lever arm, hydraulic, dead load...)	Dead load in lever arm

(continued)

(continued)

Pre-crack level:	CMOD = 0.5 mm (flexural) Mean COD = 0.2 mm (uniaxial tension)
Load level:	50% of $f_{R,p}$
Load calibration:	Load cell, before the beginning of the test
Load control in time:	No control
Transducers:	1 Pi-shaped transducer per specimen. (flexural) 3 LVDTs per specimen. (uniaxial tension)
Transducer position:	Across the notch
DAS rate: (Continuous, diary, weekly...)	Continuous
Displacements measured: (CMOD, COD, CTOD or deflection)	CMOD (flexure) COD (uniaxial tension)
Frames location: (climatic room, laboratory or outside)	Climate controlled room
Temperature controlled: (yes/no)	Yes, 21 °C ±1 °C
Humidity controlled: (yes/no)	Yes, 55% ±5%

5.10.1 Description of Creep Frames

The flexural creep test frames were designed to test a stack of up to three specimens (Fig. 5.31). They are built in steel and consist of class 2 lever arm system where a dead load is placed on the lever arm to produce the needed load on the specimens [37, 38]. This load is transferred to the top specimen by means of a steel load-plate.

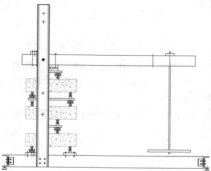

Fig. 5.31 Creep frame for bending tests

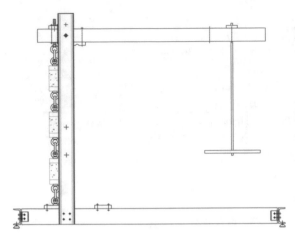 

Fig. 5.32 Creep frame for uniaxial tension tests

The specimens are supported by steel load-transmission elements which generate 4PBT load configuration. The creep frames and the DAS are located in a climatic room were temperature and humidity are controlled (T = 21 °C ± 1 °C, RH = 55% ± 5%).

The uniaxial tension creep test frames were designed to test a chain of up to three specimens (Fig. 5.32). They are built in steel and consist of class 1 lever arm system where a dead load is placed on the lever arm to produce the needed load on the specimens. Steel plates with ringbolts are glued at the ends of each cylindrical specimens. These ringbolts are connected by means of shackles in order to form a chain of specimens. The load is transferred to the specimens by attaching the top ringbolt to the lever by means of a shackle. The creep frames and the DAS are located in a climatic room were temperature and humidity are controlled (T = 21 *C ± 1 °C, RH = 55% ± 5%).

5.10.2 Support Conditions

For the flexural tests, supports are made of 20 mm diameter steel cylinders (Fig. 5.33). All these cylinders are allowed to rotate around their longitudinal axis. One of the supports of each side of the prisms also allows rotations around the longitudinal axis of the prisms in order to do not induce torsional stresses.

For the Uniaxial tension tests, the steel plates glued at the ends of each cylinder are free to rotate.

Fig. 5.33 Load transferring elements and supports

5.10.3 Measuring System and Data Acquisition

For the flexural tests, a crack opening transducer (Tokyo Measuring Instruments Lab. Type PI-5–50) was installed across the notch of each specimen at the middle of the width of the prisms. CMOD values were continuously measured using a National Instruments SCXI data acquisition system controlled by the software Labview. The load applied by each frame was measured before the beginning the tests with a load cell. The load was not measured during the tests.

For the Uniaxial tension tests, three LVDTs were installed across the notch of each specimen around their circumference. COD values were continuously measured using a National Instruments SCXI data acquisition system controlled by the software Labview. The load applied by each frame was measured before the beginning the tests with a load cell. The load was not measured during the tests.

5.10.4 Creep Test Procedure

5.10.4.1 Flexure Creep Test Procedure

Specimens were pre-cracked adopting three point-bending tests according to EN-14651. The specimens were notched at mid-span for a depth of 25 mm. The notch width was 4 mm. Tests were carried out using a MTS Landmark servo-hydraulic frame driven by a Flextest 40 controller, using closed-loop CMOD control. CMOD rates were 0.05 mm/min for CMOD < 0.1 mm and 0.2 mm/min for larger CMOD values. The CMOD was measured using a clip-on displacement transducer. When the desired CMOD of 0.5 mm was achieved, the specimens were unloaded at 0.1 mm/min. Once the specimen was unloaded, it was stored turned 90° waiting to be placed in the creep frames.

Previously to the beginning of the creep test, the specimens were stored for two days in the climate room where the creep test will be performed. During this time, the supports for the CMOD transducer to be used in the creep tests were installed. Due to the multiple specimen setup of the creep frames, the creep tests are performed in a 4PBT configuration in order to improve the stability of the column. The load to be applied in the 4PBT load configuration in computed considering the different geometry. The position of each specimen in the stack is defined based on its strength, in order of achieve the desired creep *creep index*.

Specimens are placed on the frame one-by-one. All the CMOD transducers are zeroed before putting the specimens in position. The CMOD is measured during the installation of specimens in order to measure crack openings produced by the weight of specimens and of the steel elements used to transfer loads between them. While positioning the specimens in the frame the loading lever is lifted with a hydraulic jack. Once the specimens are in position the required weight is placed on the lever and this latter is released (in about 10 s) in order to transfer the load to the stack of specimens.

Once the creep test is finished, the frames are unloaded and the specimens left in place for 30 days in order to register recovery deformations after creep test. The lever arm is gradually lifted by means of a hydraulic jack. Thus, the external load of the frame is removed and only remains the weight load of the upper specimens. After one month recording recovery deformations, the demounting process starts in a similar way than when mounting. All the specimens are stored turned 90° in order to avoid CMOD displacement due to the dead load of the specimens waiting to be tested again in flexure.

The post-creep failure test of the specimens tested in creep was performed following the EN 14651 standard recommendations. For these tests, a 3PBT load configuration is again adopted and with a load rate of 0.1 mm/min CMOD. When the CMOD achieves 4 mm, the test is finished, and the specimen is totally broken in order to have both sides of the cracked section ready to count the fibres.

5.10.4.2 Uniaxial Tension Creep Test Procedure

Cylindrical specimens were cored from $150 \times 150 \times 600$ mm beams (Fig. 5.34). The axis of the cores corresponds to the longitudinal arises of the beams in order to have a consistent direction of fibres in tension in uniaxial tension tests and flexural tests. After coring the specimens, ends were cut in order to obtain a length of 100 mm

Fig. 5.34 Extraction of the cores (left) and set-up for pre-cracking (right)

and steel plates were glued on the cut surfaces. Then a 9 mm deep circumferential notch was made at the middle of the specimens. Before creep tests specimens were pre-cracked using an original test setup in which rotations of the specimen ends a free (Fig. 5.34), in order to replicate the test conditions of long-term tests. Spherical joints were used to this aim. These joins where then connected to the hydraulic grips of the testing machine though a fork and a pin.

The pre-cracking procedure was carried out using a MTS Servo hydraulic loading frame, controlling the crack opening displacement rate with three Clip-on displacement transducers installed across the notch (120° around the circumference). In particular the following closed-loop control modes were used: while $max(COD) < 0.05$ mm a rate of 0.005 mm/min for $max(COD)$, then while $mean(COD) < 0.05$ mm a rate of 0.005 mm/min for $mean(COD)$ and finally 0.01 mm/min for $mean(COD)$ up to $mean(COD) = 0.2$ mm. The specimens were then unloaded and the residual crack opening measured. Aluminium plates were then screwed to the steel discs at the endos of each specimen in order to avoid damage while moving the specimens. Supports for the LVTDs were glued and the specimens were stored in a vertical position.

The load to be applied in the creep test was 50% of the strength measured at the end of the pre-cracking tests. The position of each specimen in the chain was defined based on its strength, in order of achieve the desired *creep index*. Specimens are placed on the frame one-by-one. All the LVDTs transducers are zeroed before putting the specimens in position. The crack opening is measured during the installation of specimens in order to measure openings produced by the weight of specimens. While positioning the specimens in the frame the loading lever is lifted up with a hydraulic jack. Once the specimens are in position the required weight is placed on the lever and this latter is released (in about 10 s) in order to transfer the load to the stack of specimens.

Once the creep test is finished, the frames are unloaded and the specimens left in place for 30 days in order to register recovery deformations after creep test. The lever arm is gradually lifted by means of a hydraulic jack. After one month recording recovery deformations, the demounting process starts in a similar way than when mounting.

The post-creep failure test of the specimens tested in creep was performed using the same setup adopted for pre-cracking. The loading rate in this case 0.1 mm/min in terms of mean (COD).

5.11 Description of Equipment/Procedures: LAB-12

Institution/company:	Universidade Federal de Rio de Janeiro
Department:	NUMATS
Information provided by:	Romildo D. Toledo Filho and Karyne Ferreira dos Santos
E-mail address:	toledo@coc.ufrj.br / karyne.ferreira@aluno.unb.br

Test method: (Flexural/Direct Tension)	Flexural and Panel
Specimens type: (Moulded, Sawed, Cores)	Moulded
N° of creep frames available:	2
N° of specimens by frame:	2
Specimens dimensions: (Prismatic, cylindrical, square panel or round panel)	Prismatic (150 × 150 × 600 mm) Square panel (600 × 600 × 100 mm)
Notch: (yes/no)	Yes (flexural beam) No (panel)
Creep Test configuration: (Only for flexural beam test: 3PBT/4PBT)	Four-point bending test (4PBT)
Creep Test configuration:	
Load application rollers: (In the case of 3PBT, only roller A)	OOX—free Rotation X, free Rotation Z, fixed Translation X XOX—fixed Rotation X, free Rotation Z, fixed Translation X
Supporting rollers: (rollers A and B freedom description)	OOX—free Rotation X, free Rotation Z, fixed Translation X XOX—fixed Rotation X, free Rotation Z, fixed Translation X
Load transfer: (lever arm, hydraulic, dead load…)	Hydraulic
Pre-crack level:	CMOD equal to 0.5 mm (flexural beam) Deflection equal to 2 mm (panel)
Load level:	50% of $f_{R,p}$ (flexural beam) 60% of F_L (panel)
Load calibration:	Pressure gauge
Load control in time:	Daily controlled and weekly adjusted
Transducers:	2 Electronic LVDT / specimen (flexural beam) 3 Electronic LVDT / specimen (panel)
Transducer position:	LVDT near the crack. In the creep frame one of the two specimens is upside-down, so it can be considered that the LVDT is up the specimen, and in the other specimen the LVDT is under the specimen

(continued)

(continued)

DAS rate: (Continuous, diary, weekly…)	Continuous
Displacements measured: (CMOD, COD, CTOD or deflection)	CMOD (flexural beam) Displacement and crack opening (panel)
Frames location: (climatic room, laboratory or outside)	Climatic room
Temperature controlled: (yes/no)	Yes, 22 °C ± 2 °C
Humidity controlled: (yes/no)	Yes, 55% ± 5%

5.11.1   Description of Creep Frames

The frames used in the flexural creep test were designed for the 4PBT load configuration operating with two specimens, as illustrated in Fig. 5.35. They were built in
steel and consist in a structure with a base [s]; four threaded bars set in the base;
and a plate [m] that ran thought by the four threaded bars, with nuts supporting the
plate [l]. The load is equally distributed by applying continuous pressure created

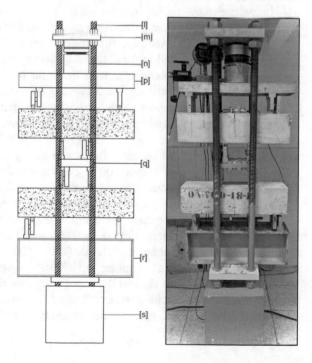

Fig. 5.35 Flexure creep
frame geometry and view of
frames during the creep test

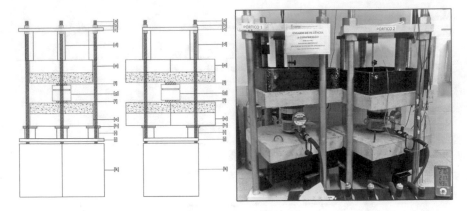

Fig. 5.36 Panel creep rig geometry and view of the rig during creep test

from an imposed displacement of a hydraulic jack [n]. The pressure is verified by a
manometer located near the hydraulic jack. There are in the frame three transmission
elements. The first one, is made of a steel H-beam of 600 mm, with two supports of
450 mm span [r]. The second [q], is positioned in the middle of the frame, between
the specimens, with two rollers up and two down, with both of 150 mm span. The
third one, is positioned above the specimen, with a length of 600 mm and span of
450 mm.

The panel creep test rig was designed to test two panels at a time, applying the
same load in both, as illustrated in Fig. 5.36. The equipment consists of a solid steel
base [k], where four threaded bars are set with two movable steel plates, one at the
top [c] and one at the bottom [j], both are passed through the bars [a] and secured by
nuts [b]. The distance between the threaded bars are of 450 mm, because of that, the
specimens had to be placed in the diagonal of the creep rig.

There were five transmission elements inside the panel creep test rig. The first [i]
were placed above the bottom steel plate [j], they consisted of four steel supports with
150 mm height, they were used to give space between the plate and the frame, so the
operator can see the cracks and place the LVDT transducers. The second transmission
element is a rigid square frame [e], of 500 × 500 × 150 mm and 20 mm thick, with
a layer of rubber between the steel frame and the specimen. The third element is
a rigid steel square [f], of 100 × 100 × 20 mm, positioned at the centre of upper
face of the specimen. Between the square and the support, there is a layer of rubber.
The fourth is the hydraulic jack positioned above the steel square, it has a steel ball
inside, leaving free rotation and adjustment. The system was symmetrical, this way,
the next elements were another square steel plaque (100 × 100 × 20 mm), rubber,
upside-down specimen, rubber and a frame (100 × 100 × 20 mm). Finally, four steel
cylinders [d] with 75 mm of diameter and 300 mm of height are used as supports.

Both the flexural creep rig and the panel creep test rig were in a climatic room
were temperature and humidity were controlled. The average temperature during the
year was of 22°C ± 2, the year average humidity was of 55% ±5. The temperature

and humidity were daily measured by a digital thermometer and hygrometer and manually recorded.

5.11.2 Support Conditions

The supports of the 4PBT for pre- and post- crack stage, also for the flexural creep test, were placed in combination, two at a time, as illustrated in Fig. 5.37. One was fixed with transversal rotation, and free longitudinal turn, as the other one was free of transversal rotation and fixed in longitudinal turn. In addition, inside the hydraulic jack, there is a steel ball that let the column to rotate and be adjusted. Hence, with these elements, the geometry errors were compensated and constrictions due to boundary conditions were avoided.

For the panel creep test simple supports were used as seen in Fig. 5.38, a rubber between the frame and the specimen was placed for better accommodation and to correct the specimen geometry errors. The hydraulic jack provided a free rotation of the column. Hence, the load is transferred in all the frame and stress concentration is avoided. In order to guarantee alignment in the flexural and panel creep apparatus, a laser line level with bubble level were used in the three directions.

Fig. 5.37 Support rollers: **a** free turn roller, **b** free rotational roller, **c** hydraulic jack with free rotation

Fig. 5.38 Support of the panel: **a** rigid frame, **b** free rotation provided of the hydraulic jack

5.11.3 Measuring System and Data Acquisition

For both systems, the flexure creep test and the panel creep test, the load in each frame is controlled by the pressure applied in the system. The manometer located near the hydraulic jack registers the pressure continuously. The pressure in the specimens was daily checked and adjusted if any variation was observed.

For the flexure creep test, it was intended to measure the CMOD in time. In order to do that, it was placed two LVDTs near the notch as seen in Fig. 5.39, and taken an average of both.

For the panel creep test, it was desired to record the displacement of the panel and crack opening of the two main cracks. In this way, each specimen was instrumented with three LVDTs as seen in Fig. 5.40. One in the centre measuring the displacement using magnetic base support fixed at the creep rig. The other two transducers were positioned in the centre, where it was supposed to be the beginning of each crack. The LVDTs were connected to a DAS, which continuously recorded the transducers readings.

Fig. 5.39 LVDTs measuring the CMOD for the flexure creep test

Fig. 5.40 LVDTs measuring the panel displacement and the crack opening of the main two cracks

5.11.4   Creep Test Procedure

The specimens were received and kept in open air for one week where temperature in that time of the year was of 28 °C and relative humidity of 62%. After this period, they were moved to the mechanical testing laboratory with temperature (23 ± 1 °C) and humidity controlled ($52 \pm 2\%$) until the pre-cracking stage.

5.11.4.1   Flexure Creep Test Procedure

The pre-cracking of the flexure creep test specimens was performed according the EN 14651 standard recommendation, however it was chosen to use the 4PBT config-uration, since the creep test was designed as a 4PBT due to stabilization of the creep frame. This way, numerical transformation was avoided.

The specimens were notched 25 mm in depth, with a saw, and moved to the mechanical testing laboratory. The 4PBT, with an upper span of 150 mm and a lower span of 450 mm, was carried out with a hydraulic universal testing machine, Shimadzu UH-F1000kN. The displacement and the CMOD were measured by LVDTs. The test rate was of 0.1 mm/min until the CMOD reached 0.1 mm and then the rate changed to 0.2 mm/min until the CMOD of 0.5 mm. When that limit was achieved, the specimen was unloaded, turned 90° and moved to the creep room.

Previously to the beginning of the flexure creep test, the creep rig was calibrated with a load cell, so the pressure applied was correlated to the wanted load. Since the creep rig was designed for two specimens, an average of the load obtained in the pre-cracking stage was calculated and applied in both specimens.

After the pre-cracking, the specimens were moved to the creep room and posi-tioned in the flexure creep frame, the LVDTs were placed in the specimens and started recording the CMOD. The specimens stayed in the creep rig for 24 h without any load, this way the specimens had time to acclimate to the environment and to record the first readings without load.

In the second day, the LVDTs were set to zero and the pressure was slowed put in. Time and CMOD were recorded while the pressure was being applied. In the beginning of the test the pressure must be reapplied every hour, since the load was due to pressure, and creep is more pronounced in the beginning of the test. After the first hours, the pressure was stabilized and checked daily and readjusted when necessary. The exact time that pressure was applied and reapplied was recorded to it could be correlated to the CMOD obtained.

The test was carried out for 360 days, thereafter, the frames were unloaded. There-unto, the pressure was slowly taken off the hydraulic jack, and the specimens stayed in the frame for more 30 days. This way, the recovery of the CMOD was measured. Following, the specimens were taken of the creep rig, rotated 90° and moved back to the mechanical testing laboratory, the LVDTs were detached when the specimens were moved.

In the mechanical testing laboratory, the specimens were tested until failure following the EN 14651 standard recommendation. The load rate was of 0.2 mm/min, and the specimen was broken in both sides so the fibres that passed the cracked section could be counted.

5.11.4.2 Square Panel Creep Test Procedure

The pre-cracking stage of the panel creep test was performed following the UNI EN 14488–5 [28] standard recommendation. The plate was supported in the four borders by a steel frame of $500 \times 500 \times 150$ mm and 20 mm of thickness. The test was carried out in a hydraulic universal testing machine, Shimadzu UH-F1000kN, and the test rate was of 1 mm/min. The displacement was measured by two LVDTs located 1 cm apart from the center of the specimen. The load was applied until a 2 mm displacement was obtained, then the specimen was unloaded and transferred to the creep laboratory by a hydraulic crane.

Before the beginning of the panel creep test, the creep rig was calibrated with a load cell, so the pressure applied by the hydraulic jack could be correlated to the wanted load. The test was prepared one step at a time; all pieces were centred and verified by laser and bubble level, to guarantee that there were no eccentricities in all three dimensions.

First, the support elements were placed [i] and [j], then the first specimen. Following, the square plate with a layer of rubber [f] was positioned in the centre of the specimen, the hydraulic jack was centred in the square plate and another square plate was placed above the hydraulic jack. The second specimen was positioned, but before it had to be turned upside down with the help of four men and a hydraulic crane. Following, another frame and the final supports [d] were placed. In order to stabilize the creep rig, it was necessary to put a small amount of pressure, less than 1% of the total load. The final step was to place the LVDTs in all the specimens and start to record the displacements. The specimens stayed in this position for 24 h, to acclimate to environmental conditions and avoid thermal interference.

After a day, the LVDTs were set to zero and the load was slowed putted on, time and displacements were recorded for 360 days. In the first hours of the test, the load had to be reapplied due to loss of pressure since the pressure was obtained by an imposed deformation, in this way when creep occurs there is a loss of pressure, which is more evidenced in the beginning of the test. When the pressure was stabilized, it was checked daily and readjusted when necessary.

Following the 360 days, the creep rig was unloaded, leaving only 1% of the pressure due to stabilization and safety. Thereunto, the pressure was slowly taken off the hydraulic jack, and the specimens stayed in the frame for more 30 days, so the recoveries of the displacements were measured.

After the 30 days of recovery, the pressure was completely taken off and the specimens were removed from the creep rig using a hydraulic crane and moved to the mechanical testing laboratory, where they were tested until complete failure following the EN 14488–5 standard recommendation. The load rate was of 1 mm/min, two LVDTs were placed 1 cm of the centre and measured the displacement. The specimens were broken in four sides so the fibres that passed the cracked section could be counted.

5.12   Description of Equipment/Procedures: LAB-13

Institution/company:	Polytechnique Montreal
Department:	Civil, Geological and Mining Engineering
Information provided by:	Jean-Philippe Charron
E-mail address:	jean-philippe.charron@polymtl.ca

Test method: (Flexural/Direct Tension)	Flexural
Specimens type: (Moulded, Sawed, Cores)	Moulded and sawed, 150 × 150 × 700 specimens were sawed in their longitudinal axis to obtain 75 × 150 × 700 mm specimens
N° of creep frames available:	1
N° of specimens by frame:	2
Specimens dimensions: (Prismatic, cylindrical, square panel or round panel)	Prismatic, 75 × 150 × 700 mm
Notch: (yes/no)	Yes
Creep Test configuration: (Only for flexural beam test: 3PBT/4PBT)	Four-point bending test (4PBT)
Load application rollers: (In the case of 3PBT, only roller A)	XOX—fixed Rotation X, free Rotation Z, fixed Translation X OXX—free Rotation X, fixed Rotation Z, fixed Translation X
Supporting rollers: (rollers A and B freedom description)	XOX—fixed Rotation X, free Rotation Z, fixed Translation X XOX—fixed Rotation X, free Rotation Z, fixed Translation X
Load transfer: (lever arm, hydraulic, dead load…)	Hydraulic, flat Freyssinet actuator loaded by a hydraulic accumulator
Pre-crack level:	CMOD equal to 0.5 mm

(continued)

(continued)

Load level:	50% of $f_{R,p}$
Load calibration:	By load cell
Load control in time:	Weekly controlled and adjusted
Transducers:	2 LVDT/specimen (displacement & crack opening)
	1 Load cell/frame
Transducer position:	LVDT on the side of the specimen (displacement)
	LVDT on the side of the specimen (crack opening)
DAS rate: (Continuous, diary, weekly…)	Continuous
Displacements measured: (CMOD, COD, CTOD or deflection)	CMOD and displacement
Frames location: (climatic room, laboratory or outside)	Climatic room
Temperature controlled: (yes/no)	Yes, 23 °C $\pm$ 2 °C
Humidity controlled: (yes/no)	Yes, 50% $\pm$ 5%

5.12.1 Description of Creep Frames

The flexural creep test frames were designed in a modified compression creep rig as seen in Fig. 5.41 where details of the frame configuration with two specimens adopted for the RRT program are presented. The frames are built in steel. The load is applied in the frame by means of a flat Freyssinet actuator, loaded by a hydraulic accumulator, located at the bottom part of the frame [39]. The swivel and the load cell are in the upper part of the frame. High-speed steel beams installed between the actuator and the load cell transfer the load to the specimens.

5.12.2 Support Conditions

Figure 5.42 shows the support types for the 4PBT configuration. There are three roller supports (fixed transversal rotation & free longitudinal turn) installed at the two supports and one loading point, and one pinned-roller support (free transversal rotation & fixed longitudinal turn) placed at one of the two loading points. The support conditions compensate geometry errors since the pinned-roller allow rotation in transversal and longitudinal axis. The thin steel plate and screws in front of the rollers in Fig. 5.42 are kept only during installation of the specimens in the frame to allow stability. Once installation of specimens is finished, plate and screws are removed to unlock displacement and rotation.

5.12.3   Measuring System and Data Acquisition

The deflection and the CMODs were measured by LVDTs on one side of the specimens (Fig. 5.43). The load was measured by a load cell in the upper part of the frame (Fig. 5.41). The signals of the LVDTs and the load cell were continuously registered by the DAS.

The load in the frames is weekly checked and if slight variations in the load are observed, the pressure was slightly modified in the flat Freyssinet actuator to adjust the load to the desired one.

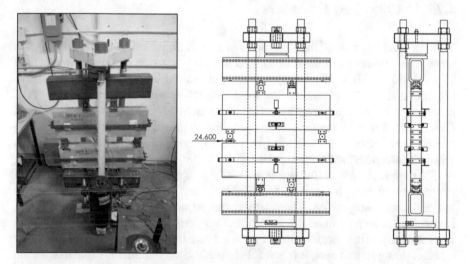

Fig. 5.41 Frame configuration with two specimens adopted for the RRT program

Fig. 5.42 Support types: **a** Pinned-roller support installed at one loading point; **b** Roller support installed at the two supports and one loading point; **c** unlocked supports

Fig. 5.43 LVDT transducers location

5.12.4 Creep Test Procedure

The pre-cracking test of the specimens was performed following RRT program recommendations. Since the frame could not accommodate larger specimens than 75 mm, 150 × 150 × 700 mm specimens were sawed in their longitudinal axis to obtain two 75 × 150 × 700 mm. The specimens where notched and tested in a 4PBT configuration in 600 mm length span at 0.5 mm/min CMOD rate. When the desired CMOD of 0.5 mm was achieved, the specimens were unloaded at 2 mm/min. Once the specimens were totally unloaded, they were stored turned 90° in the climate room where the creep test will be performed.

The creep tests are performed in a 4PBT configuration. Therefore, a numerical transformation must be realized in order to obtain the load of each specimen in a 4PBT load configuration. The relative position of each specimen in the column is decided keeping in mind the load of the upper specimens and the desired load level. Thereby, at the bottom of the column are usually placed specimens with higher residual strengths and thus, it is possible to achieve very similar *creep index* for the specimens of each frame.

The mounting of the specimens must be done step by step. Over the Freyssinet actuator is installed the first HSS steel beam. Once the first specimen is located in the frame, the LVDT transducer in connected to the DAS and the first zero is registered. After that, the supports between specimens must be located and then the second specimen installed. Since the LVDT of the bottom specimen is already connected, the CMOD due to the weight of the load transmission element and the next specimens is already being recorded. Once the second LVDT is connected, the second zero is recorded. Above the second specimen was placed chronologically the second HSS steel beam, the load cell, the swivel and the final steel plate of frame.

The load was applied from the flat Freyssinet actuator, loaded by a hydraulic accumulator, located on the bottom part of the frame. The hydraulic pressure to be applied is known from a calibration made before the creep test. The hydraulic pressure is normally increased quickly in the actuator to the expected value. Unfortunately, a problem occurred during the RRT program, the hydraulic pressure in the actuator did not achieve the expected value. It took 1.3 h to fix the problem and adjust the hydraulic pressure to the expected value corresponding to the desired loading. Due to this situation, an important part of the creep deformation occurred before the desired load was achieved in the frame. This problem of load application explains the very low creep deformation measured in the specimen tested at Polytechnique Montreal.

The load was slightly varying along time in the frame. Thus, the load cells readings were weekly checked and the load adjusted if needed in order to assure the required load in the frames during the 360 days duration of the creep test.

Once the creep tests were finished, the frame was unloaded and the specimens left in place for 30 days in order to register recovery deformations after creep test. The load of the frame was removed but it remains the weight load of the upper specimens. After one month recording recovery deformations, the demounting process starts in a similar way than when mounting. All the specimens are stored turned 90° in order to avoid CMOD displacement due to the dead load of the specimens waiting to be tested again in flexure.

The post-creep failure test of the specimens tested in creep was performed following RRT program recommendations. For these tests, a 4PBT load configuration is again adopted and the load rate is now 2 mm/min CMOD rate since the specimens are already cracked. As decided by the TC, one hysteresis loop was performed until the specific load achieved during creep test and then, the failure test continued until failure. When the CMOD achieves 4 mm, the test is finished and the specimen is totally broken in order to have both sides of the cracked section ready to count the fibres.

5.13 Description of Equipment/Procedures: LAB-15

Institution/company:	Sigma Béton
Department:	Engineering concrete department
Information provided by:	Damien Rogat
E-mail address:	damien.rogat@vicat.fr

Test method: (Flexural/Direct Tension)	Flexural panel test
Specimens type: (Moulded, Sawed, Cores)	Moulded
N° of creep frames available:	24
N° of specimens by frame:	1

(continued)

(continued)

Specimens dimensions: (Prismatic, cylindrical, square panel or round panel)	Square panel, 600 × 600 × 100 mm
Notch: (yes/no)	No
Creep Test configuration:	
Load application rollers:	Rigid steel square loading block (punch) Contact surface of 100 × 100 mm and 20 mm thickness
Supporting rollers:	Square steel frame 500 × 500 mm
Load transfer: (lever arm, hydraulic, dead load…)	Lever arm, Dead load (counterweight)
Pre-crack level:	Until the elastic limit strength, specific for each sample
Load level:	120% of elastic strength, reduced to 60% if the deflection exceeds 2 mm
Load calibration:	By load cell
Load control in time:	None, calibration at start to set the proper load and checked at the end
Transducers:	LVDT (Linear Variable Differential Transformer)
Transducer position:	Under the specimen, in the center
DAS rate: (Continuous, diary, weekly…)	Continuous, each 30 s
Displacements measured: (CMOD, COD, CTOD or deflection)	Deflection
Frames location: (climatic room, laboratory or outside)	Laboratory
Temperature controlled: (yes/no)	No Average: 20.5 °C ; Max = 28.4 °C Min = 17.2 °C
Humidity controlled: (yes/no)	No Average: 54.1°%; Max = 73.8 °C Min = 35.3 °C

5.13.1   Description of Creep Frames

The creep frames were developed during the experimental program organised by CETU (Centre for Tunnel Studies from the French Government), the association ASQUAPRO (association for the quality of sprayed concrete), the French Railway Company and Sigma Béton [40,41]. It is designed to test slabs of 60 × 60 cm with a thickness of 10 cm, same as NF EN 14488–5, as seen in Fig. 5.44.

5.13.2   Support Conditions

Figure 5.45 shows the square metallic support that is used (external dimensions 540 × 540 mm, internal dimensions 500 × 500 mm). The sample must be placed in order to have the loading block exactly at the centre. Clamping guides fix the metallic support on the creep frame.

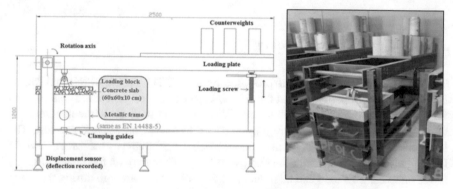

Fig. 5.44   Creep frame geometry and view of frames during creep test

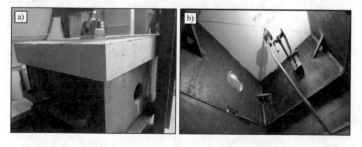

Fig. 5.45   Square panels support: **a** front view, **b** bottom view

Fig. 5.46 LVDT location
under the specimen

5.13.3　Measuring System and Data Acquisition

The LVDT is located at the bottom of each specimen as seen in Fig. 5.46. The LVDTs are connected to the DAS and record the deflection continuously.

5.13.4　Creep Test Procedure

The square panel creep test was proposed by the CETU, ASQUAPRO and Sigma Béton. It is important to note that the ASQUAPRO creep test procedure was adapted for the RRT. In the ASQUAPRO method, slabs made with the same concrete but without fibres are needed to determine the load to apply during the creep test. The load corresponds to 120% of the maximal elastic strength found by the test method NF EN 14488–5 used to measure the energy absorption capacity on the non-reinforced fibres samples. This load can be reduced to 60% of elastic strength during the creep test if a deflection of 2 mm is reached.

For the RRT, the load was determined by a punching test on four reinforced fibre concrete (Fig. 5.47). The average of maximal elastic strength was used to calculate

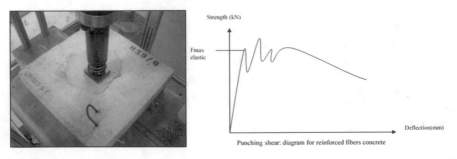

Fig. 5.47 Punching test NF EN 14488–5 and diagram

the load for the creep test. The tests to determine the load are usually performed 7 days after spraying. Since the specimens for the RRT where cast and then shipped to our facilities, the target date of seven day for the determination of the load and the pre-cracking could not be respected.

During the creep test, two loads can be applied. The first one corresponds to 120% of the maximum elastic strength (F_{max} elastic), and the second one to 60% of the maximum elastic strength. The load is reduced to 60% if a deflection of 2 mm is reached.

Calibration is performed using a load cell (Fig. 5.48). A steel slab is set in the same location as the concrete slabs to be tested. The load is adjusted by placing weights on the loading plate until the determined load is reached. The weights are marked in order to know which one needs to be removed if a deflection of 2 mm is reached.

Prior to perform the pre-cracking tests, the centre of each side of the slab is marked: above to position the punch, below to position the sensor used to measure deflection during the creep test. The sample is centred on the metallic frame. The slab is loaded with a testing machine according to the NF EN 14488–5 test method, until the first drop in the force is detected when the machine is immediately stopped. Then the slab and its metallic frame are transported to the creep test device. The LVDT sensor is installed on the mark below the sample (refer to Fig. 5.46).

The load is applied by unscrewing the loading screw under the loading plate (Fig. 5.44). The recording of deflection values begins from the start of loading. The load is maintained constant. The deflection is daily checked in the case of the deflection exceeds 2 mm. If it does, the load is reduced by 2. During the RILEM tests, none of the samples exceeded 2 mm. After one year of testing, the load is removed, the recording of deflection continues for one month after load removal. The slab is then tested according to the NF EN 14488–5 test to evaluate its residual energy absorption capacity. Specifically, for the RILEM RRT, one load/unload loop was performed until the same load applied during creep test and then loading is increase until failure.

At the end, the load value on the loading plate is checked using the same equipment as for the calibration.

Fig. 5.48 Load cell during calibration of counterweight on the creep frame

5.14 Description of Equipment/Procedures: LAB-16

Institution/company:	Stellenbosch University
Department:	Civil Engineering
Information provided by:	William P. Boshoff
E-mail address:	billy.boshoff@up.ac.za

Test method: (Flexural/Direct Tension)	Direct Tension
Specimens type: (Moulded, Sawed, Cores)	Moulded
N° of creep frames available:	Four
N° of specimens by frame:	One
Specimens dimensions: (Prismatic, cylindrical, square panel or round panel)	Prismatic, $100 \times 100 \times 500$ mm^3
Notch: (yes/no)	Yes, all around, 10 mm
Creep Test configuration:	
Load application rollers: (In the case of 3PBT, only roller A)	Not applicable
Supporting rollers: (rollers A and B freedom description)	Not applicable
Load transfer: (lever arm, hydraulic, dead load…)	Lever arm with weights
Pre-crack level:	0.2 mm crack width for SFRC and 0.14 mm for SyFRC
Load level:	50% of point of unloading during pre-cracking
Load calibration:	Load Cell
Load control in time:	Constant due to free hanging weights
Transducers:	Two LVDTs per specimen
Transducer position:	Connected to an aluminium frame over the central notch
DAS rate: (Continuous, diary, weekly…)	Continuous

(continued)

(continued)

Displacements measured: (CMOD, COD, CTOD or deflection)	COD
Frames location: (climatic room, laboratory or outside)	Climate Room
Temperature controlled: (yes/no)	Yes, 25 °C
Humidity controlled: (yes/no)	Yes 65%

5.14.1   Description of Creep Frames

The creep frames are based on a lever arm system, using free hanging weights to apply the load [12, 42]. Up to two specimens are loading in a vertical position with the two lever arms are connected in an opposite direction at the top of the column. Only one sample was tested per frame and the second sample was replaced by a chain as shown in Fig. 5.49. At the top of the column of the samples a stopper connected to a bolt was used to apply the loads on the sample by releasing the lever arms. During the test, the stopper acted as a safety mechanism, stopping the lever arms to rotate if a specimen fails.

Fig. 5.49 Creep frame showing one sample being tested

Fig. 5.50 The sample end connections and fixing mechanism in the creep frame

5.14.2   Support Conditions

This test is normally done using hooks at the ends of the samples that were cast into place. Due to the nature of this RRT, it was not possible to cast the hooks in place. Instead, steel "cups" were manufactured and glued over the ends of the specimens as shown in Fig. 5.50a. A steel plate with a hole is welded to the end of the "cup". Two steel plates with holes and bolts were used to connect the samples to the frame, Fig. 5.50b. The further connections consisted of cables and a chain was used to replace the second sample in this case. The boundary conditions were thus free to rotate at the top and bottom.

5.14.3   Measuring System and Data Acquisition

A HBM Spider8 system was used to measure the displacement using HBM LVDTs. The readings were taken once every second in the start and the rate was gradually reduced to one reading every 30 min. The two LVDTs were fixed to an aluminium frame that was fastened on the specimen over the notch as shown in Fig. 5.51.

5.14.4   Creep Test Procedure

The samples were pre-cracked in a different setup than the creep frame shown in Fig. 5.49. A temporary setup was built using an Instron hydraulic servo controlled

Fig. 5.51 The LVDT aluminium cage used to measure the crack widening over time

actuator for the purpose of these tests. The setup is shown in Fig. 5.52. The test was controlled using the CMOD measurement of one of the two LVDTs. The average crack width was monitored and the test was stopped if the average CMOD reached 0.2 mm for the SFRC and 0.14 mm for the SyFRC. After the required CMOD was reached the specimens were unloaded and carefully transported to the creep frames.

The specimens were placed in the creep frames while ensuring no external load is applied to the crack area. The lever arms were locked using the stopper at the top of the frame before the weights were applied to the arms. The stopper was released slowly (over a period of one minute) which allowed the lever arms to apply the load on the specimens.

Fig. 5.52 Testing setup used to the pre-crack the specimens

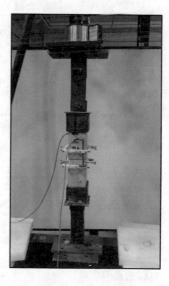

5.15 Description of Equipment/Procedures: LAB-17

Institution/company:	TSE Technologies in Structural Engineering
Department:	NA
Information provided by:	Stefan Bernard
E-mail address:	s.bernard@tse.net.au

Test method: (Flexural/Direct Tension)	ASTM C1550 Flexural testing in Round Panel
Specimens type: (Moulded, Sawed, Cores)	Moulded
Nº of creep frames available:	4
Nº of specimens by frame:	1
Specimens dimensions: (Prismatic, cylindrical, square panel or round panel)	Round Panel, 75 mm thick by 800 mm diameter
Notch: (yes/no)	No
Creep Test configuration:	
Load application rollers: (In the case of 3PBT, only roller A)	NA
Supporting conditions: (rollers/pivots freedom description)	3 support pivots with no translation, but free rotation of the panel fragments after cracking
Load transfer: (lever arm, hydraulic, dead load…)	Dead load
Pre-crack level:	2 mm deflection
Load level:	45–58% of $F_{R,p}$
Load calibration:	Calibrated load cell
Load control in time:	Load measured continuously by electronic logger
Transducers:	100 mm LVDT
Transducer position:	On loading piston
DAS rate: (Continuous, diary, weekly…)	Every 1 s, then 10 s, then 1 min, 10 min, 1 h, etc

(continued)

(continued)

Displacements measured: (CMOD, COD, CTOD or deflection)	Central deflection of loading piston
Frames location: (climatic room, laboratory or outside)	Climatic control room
Temperature controlled: (yes/no)	Yes
Humidity controlled: (yes/no)	Yes

5.15.1 Description of Creep Frames

The frames were manufactured from heavy steel plate 20–60 mm thick, very rigid (Fig. 5.53). The barrel configuration allowed the gravity loads (which comprised steel weights) to be moved up and down using a hydraulic cylinder [43, 44, 45]. However, when the load was lowered onto the test specimens the cylinder was retracted so that the loading was entirely gravity-based.

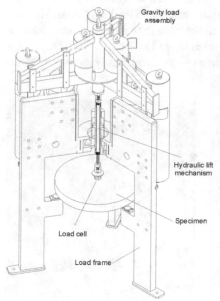

Fig. 5.53 Round panels test rig in operation

Fig. 5.54 Round panel supported on three pivoted transfer plates

5.15.2 Support Conditions

The supports consisted of three pivoted transfer plates in compliance with ASTM C1550 as seen in Fig. 5.54.

5.15.3 Measuring System and Data Acquisition

The central displacement of the specimen was measured using an LVDT positioned above the panel to assess the motion of the piston relative to the load frame. The load was measured using a calibrated DC load-cell screwed onto the end of the piston as seen in Fig. 5.55. The load cell and LVDT were logged using a DT800 data-logger from Datatek. This was programmed to log data in a logarithmically increasing series

Fig. 5.55 Close-up of LVDT
and load cell arrangement

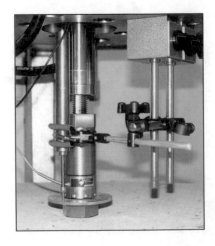

of time steps, starting once per second, then per 10 s, 1 min, 10 min, 1 h, and then every 4 h. Each 90-days tests comprised about 2000 data intervals.

5.15.4 Creep Test Procedure

The pre-cracking was performed with load frames used for the round panel test according to the ASTM C1550 in an MTS Flextest GT servo-controller driving a 100 kN MTS 244 hydraulic actuator. These tests were performed in displacement control so that the loading piston advanced at a constant rate. During the initial cracking test, this rate was 2 mm/min to provide sufficient time to stop the piston at the desired post-crack deflection of 2.0 mm. The maximum initial deflection sustained in this test was noted, then the motion of the piston was reversed at a rate of 2 mm/min and the final central deflection of the specimen noted.

The creep test stage was performed in purpose-built loading devices of the type shown in Fig. 5.53. The dimensions of the supporting fixtures were the same as a conventional ASTM C1550 [29] test configuration, but the load is applied as a gravity load consisting of steel weights rather than through hydraulic action. Each of the four test machines included a heavy steel frame to achieve high load-strain stiffness, a floating gravity load assembly with inter-changeable weights to allow the load applied in each test to be varied, and a central hydraulic actuator to raise and lower the gravity load assembly during insertion of the specimen. The load ratio target was 60%, but due to some issues with the forklift at the beginning of the creep test, the weights on the test rigs could not be changed according to the desired load. Finally, the load was as close as possible to the 60% level. The creep tests were performed over 100 days inside a climate-controlled room in which the temperature was maintained at 22.9 ± 0.5 °C, and the relative humidity was controlled at 50 ± 2%.

Upon completing the creep stage, an LVDT was placed against the underside of each panel while it still rested in the load frame depicted in Fig. 4.11. Then, the creep load was removed, and the retreat in the deflection of the centre of the panel was measured over a period of 2 h. Following completion of the unloading stage, each specimen was removed from the creep rig, transferred back to the servo-controlled testing rig, and subjected to a final ASTM C1550 [29] panel test performed at the prescribed rate of 4 mm/min piston advance to reveal the post-crack load resistance out to an additional 40 mm central deflection. This part of the testing was undertaken in the main laboratory, which was not climate-controlled but nevertheless experienced a relatively stable temperature of about 20–25 °C and relative humidity of about 60%. Given the uncertainty around the fibre counting procedure for round and square panels, the position and number of fibres in the failure surface were not assessed in these cases.

5.16 Description of Equipment/Procedures: LAB-18

Institution/company:	VSH VersuchsStollen Hagerbach AG
Department:	–
Information provided by:	Volker Wetzig / Michael Kompatscher
E-mail address:	mkompatscher@hagerbach.ch

Test method: (Flexural/Direct Tension)	Flexural panel test
Specimens type: (Moulded, Sawed, Cores)	Moulded
N° of creep frames available:	8
N° of specimens by frame:	1
Specimens dimensions: (Prismatic, cylindrical, square panel or round panel)	Square panel, 600 × 600 × 100 mm
Notch: (yes/no)	No
Creep Test configuration:	
Load application rollers:	100 × 100 mm rigid steel square loading block
Supporting rollers:	Square steel frame 500 × 500 mm
Load transfer: (lever arm, hydraulic, dead load…)	Lever arm
Pre-crack level:	Deflection equal to 3 mm
Load level:	50% of $F_{R,p}$ at 3 mm deflection
Load calibration:	–
Load control in time:	–
Transducers:	Sylvac Digital indicator S_Dial WORK BASIC
Transducer position:	Centrally on Lever arm above specimen
DAS rate: (Continuous, diary, weekly…)	1st week: twice a day 2nd to 3rd week: once per day 4th week to one year: once a week
Displacements measured: (CMOD, COD, CTOD or deflection)	Deflection

(continued)

(continued)

Frames location: (climatic room, laboratory or outside)	Laboratory
Temperature controlled: (yes/no)	Yes, 21 °C ± 2 °C
Humidity controlled: (yes/no)	Yes, 55% ± 5%

5.16.1 Description of Creep Frames

Figure 5.56 shows the test facilities used to investigate the long-term behaviour under static load. A lever system applies the required load to the test specimen [46, 47]. Test setups generate the load statically due to the weight of mass pieces placed over the lever system.

5.16.2 Support Conditions

A rigid steel box serves as a continuous support for the panels and ensures a free span of the plate of 500 mm × 500 mm (Fig. 5.57).

Fig. 5.56 View of square panel creep test frames used in VSH: **a** creep room and **b** specimen under constant loading

Fig. 5.57 Support of the
square panel

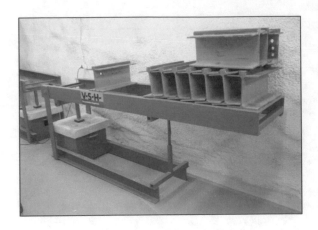

5.16.3   Measuring System and Data Acquisition

The deflection was measured with a Sylvac Digital indicator S_Dial WORK BASIC
transducer located at the top of the central arm (Fig. 5.58). Deformation values were
taken initially in short periods (half day intervals in the first week, daily in the second
week and weekly up to one year).

Fig. 5.58 Transducer type
and location

5.16.4 Creep Test Procedure

Test specimen are usually sprayed and demoulded in accordance with the requirements of EN 14488–5 [25]. The side of the specimen facing the mould base normally remained unprocessed after stripping, while all other sides were reworked by diamond sawing and grinding. Notwithstanding, for this RRT, the square panel specimens were not sprayed but poured into moulds. Therefore, no previous preparation was required.

Like in the determination of the energy absorption of steel fibre shotcrete according to EN 14488–5, the preloading of the test pieces was also carried out with deformation control. Unlike the determination of the energy absorption, the preloading stopped at a displacement of 3 mm at the centre of the plate. A displacement of 3 mm ensures that the relevant main cracks have already formed in the panel and that the tensile stresses resulting from bending are provided by the fibres bridging the cracks. After preloading, a 10×10 mm aluminium flake was glued to the underside in the centre of each plate in order to provide a secure abutment with enough repeatability after the start of the test.

The test specimens prepared in this way were stored centrally placed over the rigid steel box, which serves as a continuous support, no later than 24 h before the start of the test. The subsequent period of at least 24 h before loading was initiated to ensure that the test specimen was adapted to the climate condition in the test room.

The temperature in the test gallery (inside the mountain) is below 22 °C all year round. The heating system was able to maintain 22 °C for the duration of the test. The relative humidity was kept as constant as possible at the setpoint of 55% by regulating the fresh air supply and a humidity-controlled dehumidifier. The climate data were registered together with the deformation values.

The centric loading takes place via a ball-bearing, rigid square stamp with a side length of 100 mm. The dimensions of both the support box and the load stamp correspond to those of the testing machine for determining the energy absorption capacity according to EN 14488-5, which was also used to pre-crack the panels.

After the creep tests, the specimens were subjected no further test.

Chapter 6
Data and Parameter Definition

Aitor Llano-Torre, Pedro Serna, and Sergio H. P. Cavalaro

Abstract A convention regarding the data collection during the round-robin test (RRT) was established to simplify the analysis of results and to avoid any misunderstanding amongst laboratories. Idealised curves for each stage of the creep test are presented together with the definition of the main data to be registered during the creep test. A database was proposed for data exchange among participants. The database configuration is widely explained in the Appendix C where the parameters and coefficients requested to the RRT participants are presented, defined and explained for reference. The main parameters and variables of the RRT procedure were previously established by the participants to perform similar creep test procedures and therefore obtain comparable results. The arguments agreements of parameters such as the reference pre-crack opening level, the reference residual strength or the creep index are entirely described.

6.1 Data Collection Definition

A convention regarding the data collection during the round-robin test (RRT) was established to simplify the analysis of results. Parameters and significant values were defined and described to avoid any misunderstanding amongst laboratories. A first database proposal was presented and discussed in the RILEM TC 261-CCF. This first proposal took as reference some previous publications about database

A. Llano-Torre (✉) · P. Serna
Institute of Concrete Science and Technology ICITECH, Universitat Politècnica de València (UPV), Valencia, Spain
e-mail: aillator@upv.es

P. Serna
e-mail: pserna@cst.upv.es

S. H. P. Cavalaro
School of Architecture, Building and Civil Engineering, Loughborough University, Loughborough, UK
e-mail: s.cavalaro@lboro.ac.uk

119

A. Llano-Torre and P. Serna (eds.), *Round-Robin Test on Creep Behaviour in Cracked Sections of FRC: Experimental Program, Results and Database Analysis*, RILEM State-of-the-Art Reports 34, https://doi.org/10.1007/978-3-030-72736-9_6

on creep test [48, 49]. The RRT database configuration is widely explained in the Appendix C where the parameters and coefficients requested to the RRT participants are presented, defined and explained as reference. The final RRT database [18] was published under a Creative Commons license as supplementary material of this RRT State-of-the-Art Report and it is available for the scientific community to improve the global knowledge in the long-term deformations of the cracked fibre-reinforced concrete (FRC) specimens.

According to the agreed idealized complete curve of a creep test depicted in Fig. 4.1, in the following subsections, the subsequent idealized curves of each step of the creep test will be presented with a graphic definition and the description of the different parameters required.

It is important to highlight the importance of all these definitions and figures in order to avoid any confusing information due to poor understanding of the requested data. A significant effort to double check the information provided in the database was carried out to guarantee the validation of the collected RRT database.

Note that displacement notations are defined only for crack mouth opening displacement (CMOD). Analogous notations also apply to the crack opening displacement (COD) or deflection (δ). The analogous symbols and definitions are obtained substituting the CMOD by COD or δ.

6.1.1 Pre-cracking Stage

Figure 6.1 presents an idealisation of the stress-displacement curve and the definition of parameters obtained during the pre-crack stage.

The definitions of main parameters of the pre-cracking stage are:

f_L residual flexural tensile strength at the limit of proportionality (LOP)
$f_{R,p}$ residual flexural tensile strength at $CMOD_p$

Fig. 6.1 Pre-cracking stage parameters definition

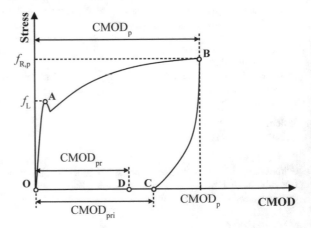

$CMOD_{pn}$ target nominal CMOD in the pre-cracking stage
$CMOD_p$ maximum CMOD reached in the pre-cracking stage
$CMOD_{pri}$ elastic CMOD recovery after unloading in the pre-cracking stage
$CMOD_{pr}$ residual CMOD 10 min after unloading in the pre-cracking stage.

6.1.2 Creep Stage

The typical stress-displacement curve during the creep stage is presented in Fig. 6.2. The definitions of parameters of the creep stage are:

$f_{R,c}$ stress applied during the creep stage
$CMOD_{ci}$ instantaneous CMOD immediately after reaching the reference load
$CMOD_{cd}^j$ delayed CMOD after j days in the creep test
$CMOD_{ct}^j$ total CMOD after j days in the creep test.

Due to the difficulty to reach a consensus about the definition of instantaneous deformation $CMOD_{ci}$ after loading, multiple short-term reference values were registered to check the influence of this definition criterion in the creep coefficients. As shown in Fig. 6.3, three different values were obtained:

t_{ci} time duration of the loading process in the creep test
$CMOD_{ci}$ instantaneous CMOD deformation immediately after reaching the reference load
$CMOD_{ci}^{10'}$ short-term CMOD deformation 10 min after reaching the reference load
$CMOD_{ci}^{30'}$ short-term CMOD deformation 30 min after reaching the reference load.

Once the creep test duration time is reached, the specimens were unloaded but not removed from the creep frames for 30 days, recording the elastic and the delayed

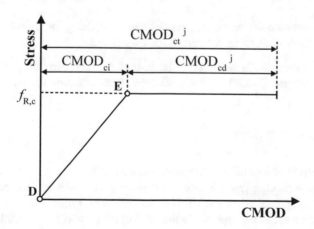

Fig. 6.2 Creep stage parameters

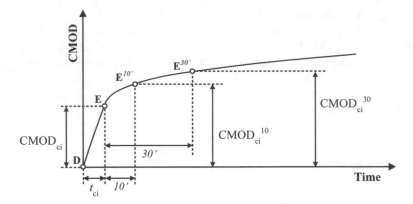

Fig. 6.3 Instantaneous and short-term deformation parameters

Fig. 6.4 Unloading stage
parameters

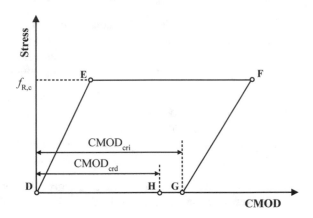

recovery deformations. The parameters defined concerning the unloading stage are
depicted in Fig. 6.4 and were defined as follows:

$CMOD_{cri}$ elastic CMOD recovery after unloading the creep test
$CMOD_{crd}$ delayed CMOD recovery 30 days after unloading the creep test.

6.1.3 Post-creep Stage

The post-creep failure bending tests were performed following the same procedure
than in the pre-cracking test, performing one hysteresis loop as defined in Fig. 4.4
and deemed completed when a CMOD deformation of 4 mm is achieved. Then the
specimens were unloaded and no further deformation was recorded. In this stage
were not obtained any significant parameter since the post-creep residual parameters
must be obtained when the three stages are ensembled.

6.1.4 Complete Creep Test Curve

Once the curves of three main stages (pre-cracking, creep and post-creep) are assembled, the post–creep residual strength parameters can be obtained at $CMOD_2$, $CMOD_3$ and $CMOD_4$ as illustrated in Fig. 6.5 referred to the origin of deformations O. If the level of deformation corresponding to $CMOD_2$, $CMOD_3$ or $CMOD_4$ is reached during the creep stage, the corresponding residual strength was not considered.

$f_{PostCreep,R,2}$ residual flexural tensile strength corresponding to origin at $CMOD_2$
$f_{PostCreep,R,3}$ residual flexural tensile strength corresponding to origin at $CMOD_3$
$f_{PostCreep,R,4}$ residual flexural tensile strength corresponding to origin at $CMOD_4$

6.2 Round-Robin Creep Test Parameter Definition

Although the creep methodologies available in the literature are similar, there are some variations in the basic parameters that could lead to a difficult comparison of results between them. In order to perform similar creep test procedures and therefore comparable results, the main parameters and variables of the RRT procedure were previously established by the participants. Thus, a better comparison of the methodological influences and the variability of the main parameters could be performed, as was established as main goal of the RRT.

Nevertheless, it was also considered important to define these parameters in a level that should be representative of the actual working conditions in a real structure in service conditions. The parameters were discussed during a RILEM TC 261-CCF meeting, even than the agreement was not always unanimous.

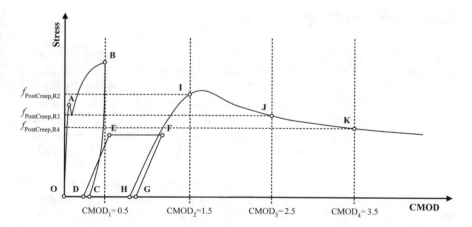

Fig. 6.5 Post-creep stage parameters

The following sections present the parameters and variables established in the RRT. These were recorded by each lab to explore their influence on the results obtained during the creep test.

6.2.1 Reference Pre-crack Opening Level

In most structures, the time-dependent deformations are related to the sustained load applied during the service stage, which should comply with crack-opening requirements of Service Limit State (SLS). The use in the RRT of cracks bigger than the allowed in SLS would not be representative of common practice as structural elements would hardly ever be allowed to remain for extended periods in such conditions. Consequently, the reference crack opening defined at the pre-cracking stage should be linked to the allowed opening in SLS.

By this token, the reference pre-crack opening level for the *flexural creep test* of prisms was related to a CMOD of 0.5 mm, which is usually taken as a reference in the literature for the evaluation of the SLS [1]. In the *direct tension tests*, it was defined a COD of 0.2 mm as the pre-cracking level. The latter is acceptable in most structural codes and is equivalent to the average crack opening found in the flexural tests of prisms for a CMOD of 0.5 mm considering a variable crack opening along the cross-section. For *round* and *square panel tests*, the crack opening at the pre-cracking stage was controlled indirectly by the deflection measured at the centre of the panel following the previous experience of researchers in this field [10, 11]. Therefore, a deflection value of 2 mm was defined for all panel tests.

6.2.2 Reference Residual Strength

The structural contribution of fibres is normally evaluated through the results of the flexural test according to EN14651 [22]. The test provides the residual flexural strengths ($f_{R,1}$ and $f_{R,3}$) used in the Model Code 2010 [1] to predict the parameters f_{Fts} and f_{Ftu} for the tensile behaviour models (rigid-plastic or linear) employed in the design of structures. In the simplified "*Rigid-plastic model*", this f_{Ftu} value equals $f_{R,3}/3$. Note that, with this criterion, both concrete mixes characterised in the RRT have similar f_{Ftu} performance (1.11 and 1.15 MPa for SyFRC and SFRC respectively).

The "*Linear model*" is convenient if a complete structural analysis is performed. It considers the tensile behaviour depending on the ultimate crack opening when analysing the Ultimate Limit State (ULS), being also suitable if fibres are used for crack control. It is usually taken as a reference to define the residual strength on creep studies in the bibliography [9,15]. For SLS crack openings, the "*Linear model*" considers that f_{Fts} equals $0.45 f_{R,1}$.

Since the RRT intends to simulate the cracking and loading in SLS, the reference stress level during the flexural creep stage of prisms was defined as the 50% of $f_{R,p}$ at $CMOD_p = 0.5$ what is equal to 50% of $f_{R,1}$. With this criterion, the reference equivalent residual strength applied during the creep stage changes considerably depending on the fibre material used in the RRT (0.98 MPa for SyFRC mixes and 1.44 MPa for SFRC mixes).

6.2.3 Specimen or Batch Average Reference Load

The sustained load applied during the creep stage is generally taken as a percentage of the stress for a particular crack opening or deflection in the pre-cracking stage. The calculation may be made at an individual specimen level, batch level or even for each FRC level. If an individual specimen level is adopted, the stress ratio is maintained constant but the actual load applied in the creep stage varies from one specimen to the other following the scatter in their residual strength. Alternatively, the reference load may be adopted as a batch level by multiplying the average residual strength measured for all specimens in the batch and the stress ratio. In this last case, although the load for all specimens of the same batch during the creep stage is the same, the stress applied or creep index (I_c) varies from one specimen to the other and consequently, a higher scatter should be expected.

It was considered that the proposal of constant stress ratio observed across the specimens in the first approach could lead to smaller scatter on the results. By contrast, results would be derived about the nominal behaviour of the batch in the second approach. The first approach was adopted after thorough consideration to be consistent with the majority of studies from the literature and to favour the comparison of the results from different laboratories. Note that, when a multi-specimen setup was adopted, specimens were usually grouped according to their residual strength in the pre-cracking stage to minimise the variation in the stress ratio amongst multiple specimens in the same frame.

6.2.4 Creep Index (I_c) or Stress Level

It is usually assumed that creep in compression should be proportional to the stress level applied from 0% to 40% of concrete compressive strength. In standard compressive creep tests, a common stress level ranges from 30 to 35% of f_c. This range can be related to an expected stress level if the common safety coefficients are considered in the design of real structures. The stress level adopted for the complementary creep in compression tests was 40% of f_c (see Sect. 3.4).

In the case of creep tests in cracked state, different stress levels were adopted depending on the methodology. Based on discussions held in the RILEM TC 261-CCF, the agreed sustained nominal stress level or *nominal creep index* of stress (I_n)

defined for the *flexural creep test* and *direct tension creep test* for the RRT was 50% of $f_{R,p}$ at CMOD = 0.5 mm and COD = 0.2 mm respectively, even if it is more frequent lower levels in real service conditions, but reducing this level could lead to very low creep deformations more difficult to be analysed. Therefore, the applied stress level or *applied creep index* of stress (I_c) is defined as the ratio between the sustained stress during creep stage ($f_{R,c}$) and the residual strength at $CMOD_p$ ($f_{R,p}$) during the pre-cracking stage. In the case of the both *square and round panel creep tests*, the *nominal creep index* I_n was defined as 60% of F_L, defined as the ratio between the sustained load during creep stage ($F_{R,c}$) and the load at the Limit of Proportionality LOP (F_L). These values are considered representative of structures under service conditions, agreeing with the stress level used in several studies [4].

Chapter 7
Round-Robin Test Results

Aitor Llano-Torre and Pedro Serna

Abstract The results obtained from this round-robin test (RRT) on creep in fibre-reinforced concrete (FRC) specimens were grouped into a large database. More than 140 parameters and variables were registered for each FRC specimen during the RRT and compiled in an Excel file. This chapter presents only a part of the results obtained by each participant in the creep tests. The values obtained by each specimen of the most significant parameters of both pre-cracking and creep stages together with the evolution in time of delayed deformation figures are presented and analysed for each participant laboratory. The RRT creep database is available open access to the scientific community to improve the global knowledge in the long-term deformations of the cracked FRC specimens.

The round-robin test (RRT) results are arranged in four sections depending on the creep test methodology. For each participant laboratory and methodology, a short table with some significant data is given as well as the total deformations $CMOD_{ct}$, COD_{ct} or δ_{ct} registered during the creep tests to get an approach of the magnitude of the deformations obtained by participant. Note that the residual deformations from pre-cracking test ($CMOD_{pr}$, COD_{pr} or δ_{pr}) are not included. The figures represent in both linear and logarithmic time axis the total delayed deformations values given in the database for both macro-synthetic fibre-reinforced concrete (SyFRC) and steel fibre-reinforced concrete (SFRC) mixes at certain time lapses. The defined time lapse for long-term data exchange was 30 days, even if during the first 14 days additional readings were required at 1, 2, 3, 5, 7 and 14 days since most of deformations occurs during first days of creep test. This procedure was agreed to make easier the exchange of data even if some values between the defined time lapse readings were lost.

A. Llano-Torre (✉) · P. Serna
Institute of Concrete Science and Technology ICITECH, Universitat Politècnica de València (UPV), Valencia, Spain
e-mail: aillator@upv.es

P. Serna
e-mail: pserna@cst.upv.es

© RILEM 2021

A. Llano-Torre and P. Serna (eds.), *Round-Robin Test on Creep Behaviour in Cracked Sections of FRC: Experimental Program, Results and Database Analysis*, RILEM State-of-the-Art Reports 34, https://doi.org/10.1007/978-3-030-72736-9_7

127

7.1 Flexural Tests

Flexural creep tests were performed by 12 laboratories and a total of 86 prismatic specimens were tested in flexural creep as shown in Table 4.1. A brief introduction of the results obtained in flexural tests will be given for each participant.

The LAB-01 participant tested in flexural creep a total of 12 specimens: six of each FRC mix. Data obtained from pre-cracking tests as well as the stress applied and the creep index I_c during creep test for each specimen are exposed in Table 7.1. The deformations registered during creep test are represented in Fig. 7.1. Unfortunately,

Table 7.1 Main results obtained from pre-cracking tests and creep test by LAB-01

Specimen	f_L (MPa)	$f_{R,p}$ (MPa)	$CMOD_p$ (μm)	$CMOD_{pri}$ (μm)	$CMOD_{pr}$ (μm)	$f_{R,c}$ (MPa)	I_c (%)
M-B1-051	3.98	2.51	516.8	286.4	241.3	1.28	50.95
M-B1-052	3.75	2.19	535.4	326.3	288.5	1.09	50.03
M-B1-053	3.98	2.40	527.3	292.3	252.7	1.18	49.02
M-B2-159	4.07	2.21	519.4	271.1	233.7	1.11	50.45
M-B2-160	4.25	1.77	523.1	308.5	253.7	0.90	50.65
M-B2-162	4.05	2.04	529.3	301.6	267.8	0.99	48.58
S-B1-272	3.94	2.39	524.3	351.8	338.6	1.21	50.44
S-B1-273	4.26	1.96	517.8	370.6	363.3	0.99	50.38
S-B1-274	4.04	2.24	520.3	340.1	323.0	1.10	48.94
S-B2-363	4.24	2.26	512.0	341.2	312.4	1.16	51.42
S-B2-364	4.00	2.47	528.8	344.3	322.8	1.27	51.29
S-B2-366	4.23	2.23	524.3	335.6	315.7	1.05	46.79

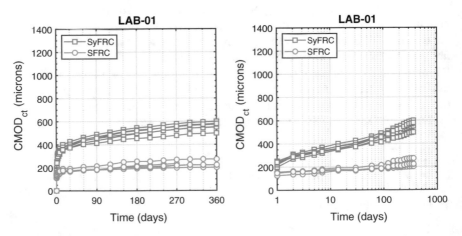

Fig. 7.1 CMOD-time curves obtained in flexural creep test by LAB-01

Fig. 7.2 Temperature and
relative humidity registered
during creep test by LAB-01

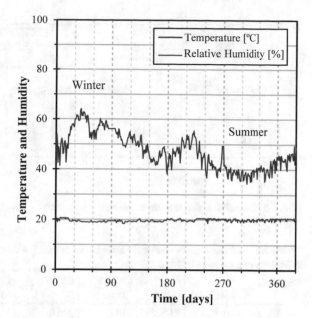

one SFRC specimen transducer failed and there is only data of 5 specimens. As it can
be seen, SyFRC and SFRC mixes are clearly differentiated as expected. The scatter
of the results obtained by LAB-01 is quite fitted and therefore the results are quite
robust. As reference, the delayed deformations developed by SyFRC specimens are
around 2.5 times higher than the SFRC specimens.

Moreover, the evolution in time of the environmental conditions in the climate
chamber of the LAB-01 are depicted in Fig. 7.2. This representation can be very
useful in the case of sudden or atypical deformations since, in some cases, they can
be explained due to changes in the environmental conditions.

After 360 under sustained loading and 30 additional days unloaded for delayed
recovery deformation, specimens were tested in flexure until failure following the
procedure described in Sect. 6.1.3. Once the three stages were performed, the curves
of the whole process of the creep test could be obtained by joining all the stages
to assess if the residual performance was influenced by the sustained loading creep
tests. The complete curves for both the SyFRC and SFRC specimens tested by LAB-
01 are represented in Fig. 7.3 including the post-creep residual strengths for each
specimen.

The LAB-02 participant tested in creep a total of 12 specimens: 6 of each fibre
material. Data obtained from pre-cracking tests as well as the stress applied and the
creep index I_c during creep test for each specimen are exposed in Table 7.2. The
deformations registered during creep test by LAB-02 are represented in Fig. 7.4.
As it can be seen in the figure, due to some issues in DAS equipment, the readings
from 90 to 300 days were lost. Fortunately, the signal was recovered for the final
weeks of the RRT for most specimens. This laboratory presents higher scatter for

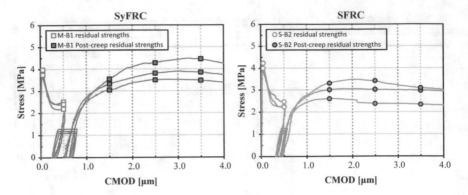

Fig. 7.3 Complete creep test curves including post-creep residual strengths for LAB-01

Table 7.2 Main results obtained from flexural pre-cracking tests and creep test by LAB-02, where CMOD is expressed in microns

Specimen	f_L (MPa)	$f_{R,p}$ (MPa)	$CMOD_p$ (μm)	$CMOD_{pri}$ (μm)	$CMOD_{pr}$ (μm)	$f_{R,c}$ (MPa)	I_c (%)
M-B1-056	4.73	2.95	513.0	236.0	–	1.51	51.19
M-B1-057	4.34	2.51	513.0	250.0	–	1.25	49.80
M-B1-058	4.21	2.08	513.0	173.0	–	1.07	51.44
M-B2-164	4.38	1.88	500.0	221.0	–	0.93	49.71
M-B2-165	4.83	2.07	514.0	337.0	–	1.00	48.41
M-B2-166	4.73	2.25	518.0	295.0	–	1.10	49.11
S-B1-277	4.29	2.41	499.0	–	–	1.18	48.90
S-B1-278	4.19	2.07	520.0	302.0	–	1.05	50.65
S-B1-279	4.38	2.57	478.0	–	–	1.31	50.79
S-B2-368	4.09	2.37	521.0	–	–	1.18	49.59
S-B2-370	4.42	2.64	519.0	299.0	–	1.46	55.42
S-B2-372	4.05	3.74	482.0	–	–	1.59	42.43

SyFRC than for SFRC. It can be observed that one of the SFRC specimens (S-B2-370) develops higher deformations than the mean average which can be explained because of the 55.4% of creep index applied for that specimen, the highest creep index of the LAB-02.

The LAB-03 participant tested a total of 6 FRC specimens, 3 of each fibre material, and the deformations registered during creep test are represented in Fig. 7.5. Data obtained from pre-cracking tests as well as the stress applied and the creep index I_c during creep test for each specimen are exposed in Table 7.3. This laboratory presents a very low scatter referring to the creep index I_c due to the single frame testing setup. SFRC specimens delayed deformations tend to decrease in time as observed in Fig. 7.5. In addition, the LAB-03 participant is the only one that kept the

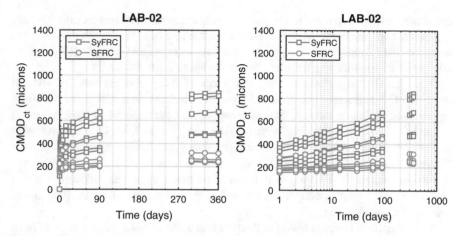

Fig. 7.4 CMOD-time curves obtained in flexural creep test by LAB-02

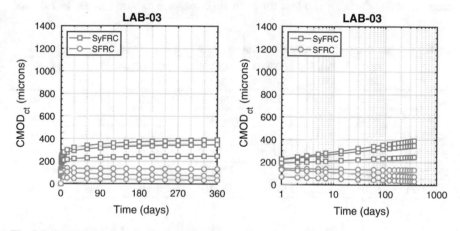

Fig. 7.5 CMOD-time curves obtained in flexural creep test by LAB-03

Table 7.3 Main results obtained from flexural pre-cracking tests and creep test by LAB-03

Specimen	f_L (MPa)	$f_{R,p}$ (MPa)	δ_p (μm)	δ_{pri} (μm)	δ_{pr} (μm)	$f_{R,c}$ (MPa)	I_c (%)
M-B1-060	5.21	2.82	500.0	276.0	–	1.41	50.00
M-B1-062	4.96	2.77	500.0	268.0	–	1.39	50.18
M-B1-063	5.09	3.13	500.0	279.0	–	1.56	49.84
S-B1-283	5.31	3.29	500.0	349.0	–	1.66	50.46
S-B1-284	5.07	2.60	500.0	376.0	–	1.30	50.00
S-B1-285	5.07	3.43	500.0	348.0	–	1.71	49.85

specimens wrapped in aluminium sheets during creep test and this fact may reduce the basic shrinkage deformations. Note that in the case of LAB-03, the results given were obtained from a fitted line to the raw data as explained in Sect. 5.3.4, where additional information can be found about the specific procedure.

The LAB-04 participant tested in creep a total of six specimens: three SyFRC and three SFRC specimens. Data obtained from pre-cracking tests as well as the stress applied and the creep index I_c during creep test for each specimen are exposed in Table 7.4. The deformations registered during creep test by LAB-04 are represented in Fig. 7.6. Both groups are clearly differentiated and SyFRC specimens present higher scatter than SFRC specimens as it can be seen in the figure. There is one SyFRC specimen (M-B1-065) which develops much more delayed deformations since, due to the multi-specimen set-up, that specimen was tested under 10% higher creep index than the rest.

The LAB-05 participant tested in flexural creep a total of 12 specimens: 6 of each FRC. Data obtained from pre-cracking tests as well as the stress applied and the creep index I_c during creep test for each specimen are exposed in Table 7.5. The

Table 7.4 Main results obtained from flexural pre-cracking tests and creep test by LAB-04

Specimen	f_L (MPa)	$f_{R,p}$ (MPa)	$CMOD_p$ (μm)	$CMOD_{pri}$ (μm)	$CMOD_{pr}$ (μm)	$f_{R,c}$ (MPa)	I_c (%)
M-B1-065	4.63	2.50	510.0	266.0	242.0	1.31	52.40
M-B1-074	4.48	2.99	510.0	255.0	237.0	1.42	47.49
M-B2-168	5.09	2.41	510.0	290.0	250.0	1.17	48.55
S-B1-290	3.99	2.97	510.0	306.0	301.0	1.65	55.56
S-B1-292	4.74	3.59	512.0	304.0	284.0	1.74	48.47
S-B2-381	5.20	3.85	510.0	320.0	294.0	1.85	48.05

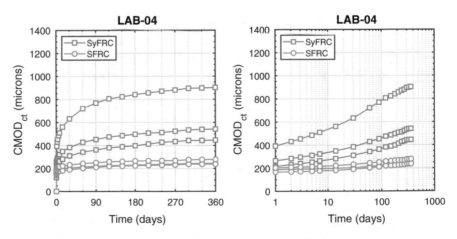

Fig. 7.6 CMOD-time curves obtained in flexural creep test by LAB-04

Table 7.5 Main results obtained from flexural pre-cracking tests and creep test by LAB-05

Specimen	f_L (MPa)	$f_{R,p}$ (MPa)	$CMOD_p$ (μm)	$CMOD_{pri}$ (μm)	$CMOD_{pr}$ (μm)	$f_{R,c}$ (MPa)	I_c (%)
M-B1-076	4.47	2.76	498.1	200.7	200.7	1.43	51.75
M-B1-077	4.68	2.81	501.5	211.9	211.9	1.35	48.04
M-B1-079	4.91	2.81	496.8	216.0	213.8	1.41	50.19
M-B2-177	5.51	2.49	496.5	194.0	191.6	1.22	48.94
M-B2-178	5.57	2.50	495.9	214.0	212.6	1.26	50.41
M-B2-179	5.69	2.65	491.5	211.0	207.2	1.34	50.56
S-B1-296	4.91	3.33	492.8	312.0	305.9	1.64	49.28
S-B1-297	4.72	3.32	491.2	266.0	261.8	1.68	50.59
S-B1-298	4.75	3.02	487.7	304.0	296.8	1.52	50.41
S-B2-431	5.78	3.21	496.0	269.0	267.0	1.65	51.38
S-B2-432	5.27	3.16	491.5	283.0	277.8	1.56	49.43
S-B2-433	5.68	3.02	487.9	301.0	293.1	1.49	49.29

deformations registered during creep test by LAB-05 are represented in Fig. 7.7 and both SyFRC and SFRC mixes are quite grouped. The long-term deformations of SyFRC specimens are around three times higher than the SFRC. A slight ascendant trend can be observed after 360 days of creep test, what could be understood as tertiary creep, but none of the specimens failed. It should be reminded that this participant started the creep test four months later than the rest due to custom issues.

The LAB-06 participant tested in flexural creep a total of six specimens: two SyFRC and four SFRC specimens. Data obtained from pre-cracking tests as well as the stress applied and the creep index I_c during creep test for each specimen are exposed in Table 7.6. The deformations registered during the creep test by LAB-06

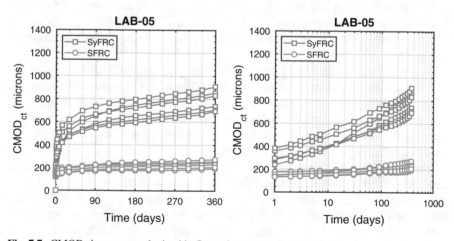

Fig. 7.7 CMOD-time curves obtained in flexural creep test by LAB-05

Table 7.6 Main results obtained from flexural pre-cracking tests and creep test by LAB-06

Specimen	f_L (MPa)	$f_{R,p}$ (MPa)	$CMOD_p$ (μm)	$CMOD_{pri}$ (μm)	$CMOD_{pr}$ (μm)	$f_{R,c}$ (MPa)	I_c (%)
M-B1-080	4.25	2.76	500.0	310.0	250.0	1.50	54.35
M-B1-082	4.38	2.43	500.0	288.0	248.0	1.29	53.09
S-B1-300	4.82	3.56	500.0	312.0	285.0	1.70	47.75
S-B1-301	4.11	2.27	500.0	360.0	305.0	1.48	65.20
S-B0-434	3.78	3.47	500.0	300.0	240.0	1.90	54.76
S-B0-435	3.95	2.67	500.0	291.0	271.0	1.68	62.92

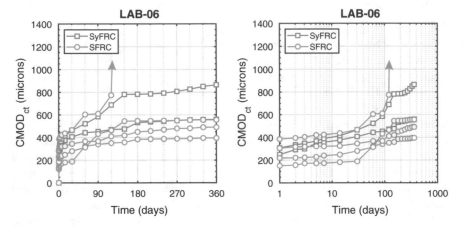

Fig. 7.8 CMOD-time curves obtained in flexural creep test by LAB-06

are represented in Fig. 7.8. One of the specimens (S-B1-301) failed at 120 days of creep test with an applied creep index of 65.2%, the highest of the specimens of LAB-06. The scatter is not only significant in terms of delayed displacement, but also regarding the instantaneous deformations of the SFRC specimens as observed in Fig. 7.8b.

In general terms, the higher scatter observed in delayed CMOD displacement curves correspond to those participants with higher scatter in terms of creep index. As previously commented on Sect. 4.1.2, the multi-specimen set-up is one of the main reasons of this scatter on creep index parameter. The participant LAB-06 presents for the SFRC specimens the highest mean creep index of this RRT with 57.66% and a very high Coefficient of Variation (CV) of 13.8%.

The LAB-07 participant tested in flexural creep a total of 6 specimens: 3 of each fibre material, and the delayed deformations registered during creep test are represented in Fig. 7.9. Data obtained from pre-cracking tests as well as the stress applied and the creep index I_c during creep test for each specimen are exposed in Table 7.7. Both materials behaviours are clearly visible in the figure, where SyFRC specimens develop deformations near 4 times higher than the SFRC specimens.

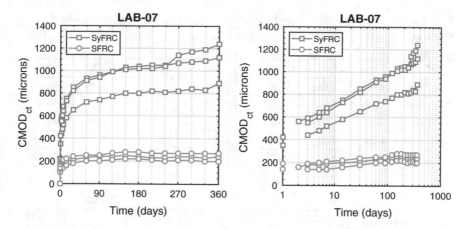

Fig. 7.9 CMOD-time curves obtained in flexural creep test by LAB-07

Table 7.7 Main results obtained from flexural pre-cracking tests and creep test by LAB-07

Specimen	f_L (MPa)	$f_{R,p}$ (MPa)	δ_p (μm)	δ_{pri} (μm)	δ_{pr} (μm)	$f_{R,c}$ (MPa)	I_c (%)
M-B2-182	4.36	2.58	465.0	–	–	1.29	50.00
M-B2-183	4.23	3.42	465.0	–	–	1.71	50.00
M-B2-184	4.26	2.48	465.0	–	–	1.22	49.19
S-B0-439	4.64	2.79	465.0	–	–	1.40	50.18
S-B0-440	4.94	4.26	465.0	–	–	2.13	50.00
S-B0-441	4.64	5.16	465.0	–	–	2.58	50.00

Single specimen set-up was adopted during creep tests and the mean creep index I_c was 49.9% with a 0.7% CV.

The LAB-08 participant tested in flexural creep a total of 6 specimens, 2 SyFRC and 4 SFRC specimens in three creep frames with a multi-specimen configuration of two specimens on each. Data obtained from pre-cracking tests as well as the stress applied and the creep index I_c during creep test for each specimen are exposed in Table 7.8. The mean creep index I_c applied on these specimens was 49.75% with a 9.7% CV. The deformations registered during creep test are represented in Fig. 7.10 where it can be observed that both SyFRC and SFRC mixes are grouped with slightly higher delayed deformations for SyFRC specimens. Regarding the instantaneous deformations when applying the load in the frame, all the specimens develop similar values. The evolution in time of the CMOD deformation seems to be quite stabilized for SFRC specimens, whereas for SyFRC specimens, is still increasing.

The LAB-10 participant tested in flexural creep in individual frames a total of 8 specimens, 4 of each fibre material, with an applied mean creep index I_c of 50.70% and a negligible CV of 0.5%. Data obtained from pre-cracking tests as well as the stress applied and the creep index I_c during creep test for each specimen are exposed

Table 7.8 Main results obtained from flexural pre-cracking tests and creep test by LAB-08

Specimen	f_L (MPa)	$f_{R,p}$ (MPa)	$CMOD_p$ (μm)	$CMOD_{pri}$ (μm)	$CMOD_{pr}$ (μm)	$f_{R,c}$ (MPa)	I_c (%)
M-B2-188	4.40	2.96	500.0	250.0	–	1.36	45.94
M-B2-189	4.32	2.58	500.0	250.0	–	1.36	52.71
S-B0-444	4.39	4.20	500.0	280.0	–	2.29	54.52
S-B0-445	4.12	4.38	500.0	270.0	–	2.14	48.93
S-B0-448	4.24	4.25	500.0	250.0	–	2.29	53.88
S-B0-449	4.45	5.04	500.0	280.0	–	2.14	42.52

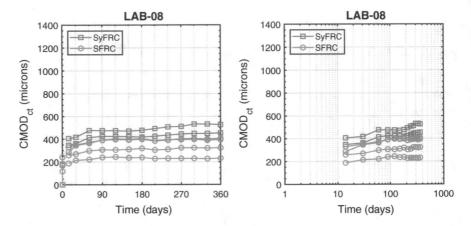

Fig. 7.10 CMOD-time curves obtained in flexural creep test by LAB-08

in Table 7.9. The CMOD deformations registered during creep test are represented in Fig. 7.11. Unfortunately, as it can be seen in the figure, certain readings were lost for both mixes due to some issues in the DAS unit. Despite this incidence, both

Table 7.9 Main results obtained from flexural pre-cracking tests and creep test by LAB-10

Specimen	f_L (MPa)	$f_{R,p}$ (MPa)	$CMOD_p$ (μm)	$CMOD_{pri}$ (μm)	$CMOD_{pr}$ (μm)	$f_{R,c}$ (MPa)	I_c (%)
M-B2-191	4.88	3.30	514.0	234.0	215.0	1.66	50.42
M-B2-192	4.71	3.05	505.0	248.0	218.0	1.54	50.58
M-B2-193	4.67	2.25	504.0	260.0	233.0	1.16	51.28
M-B2-194	5.47	2.98	505.0	249.0	225.0	1.51	50.65
S-B0-450	5.11	5.08	508.0	297.0	290.0	2.58	50.79
S-B0-451	4.40	3.68	505.0	299.0	289.0	1.86	50.65
S-B0-452	4.77	3.83	503.0	302.0	296.0	1.94	50.59
S-B0-453	4.92	4.44	503.0	312.0	303.0	2.25	50.65

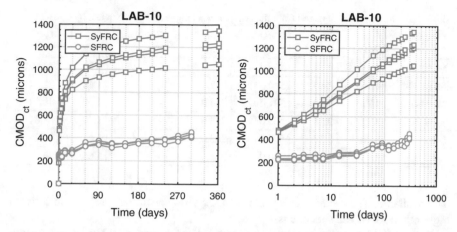

Fig. 7.11 CMOD-time curves obtained in flexural creep test by LAB-10

groups are very distinguishable with a significant scatter for SyFRC specimens. The delayed deformations for SyFRC specimens are around 3 times higher than for SFRC specimens.

The LAB-11 participant tested in flexural creep a total of 6 specimens, 3 of each fibre material, in a multi-specimen set-up with three specimens by frame where a mean creep index I_c of 53.02% was applied with a significant CV of 14.7%. This high CV in creep index parameter is given by only one specimen (S-B1-305) that is under a 68.81% creep index, and therefore develops higher delayed deformations as it can be seen in Fig. 7.12. Data obtained from pre-cracking tests as well as the stress applied and the creep index I_c during creep test for each specimen are exposed in Table 7.10. The transducer located in the specimen M-B1-092 failed and there is no

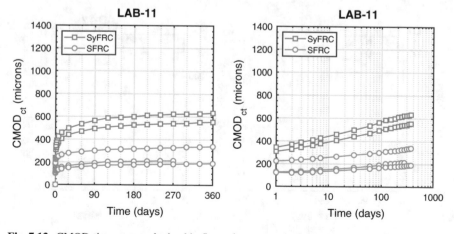

Fig. 7.12 CMOD-time curves obtained in flexural creep test by LAB-11

Table 7.10 Main results obtained from flexural pre-cracking tests and creep test by LAB-11

Specimen	f_L (MPa)	$f_{R,p}$ (MPa)	$CMOD_p$ (μm)	$CMOD_{pri}$ (μm)	$CMOD_{pr}$ (μm)	$f_{R,c}$ (MPa)	I_c (%)
M-B1-085	4.13	2.62	500.0	243.0	–	1.30	49.62
M-B1-087	4.16	2.69	500.0	277.0	–	1.40	52.04
M-B1-092	4.21	2.46	501.0	214.0	–	1.20	48.78
S-B1-305	3.71	2.18	500.0	315.0	–	1.50	68.81
S-B1-308	3.56	3.27	500.0	293.0	–	1.61	49.24
S-B1-310	3.98	3.35	500.0	346.0	–	1.66	49.63

data available. The delayed deformations for SyRC specimens are around 3 times higher than for SFRC specimens.

The LAB-12 participant tested in flexural creep a total of 4 specimens in two creep frames in a double-specimen set-up: one frame for two SyFRC specimens and another frame for two SFRC specimens. Data obtained from pre-cracking tests as well as the stress applied and the creep index I_c during creep test for each specimen are exposed in Table 7.11. As it can be observed in Fig. 7.13, there is a significant

Table 7.11 Main results obtained from flexural pre-cracking tests and creep test by LAB-12

Specimen	f_L (MPa)	$f_{R,p}$ (MPa)	$CMOD_p$ (μm)	$CMOD_{pri}$ (μm)	$CMOD_{pr}$ (μm)	$f_{R,c}$ (MPa)	I_c (%)
M-B1-095	4.69	3.35	496.0	–	–	1.75	52.24
M-B1-097	4.71	2.93	571.0	–	–	1.81	61.88
S-B1-313	4.72	4.88	536.0	–	–	2.33	47.73
S-B1-316	5.00	3.95	537.0	–	–	2.39	60.57

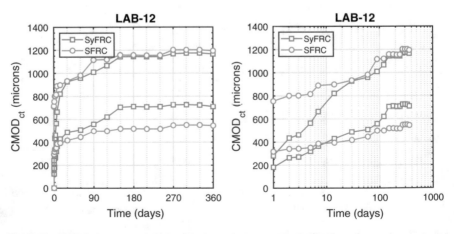

Fig. 7.13 CMOD-time curves obtained in flexural creep test by LAB-12

Table 7.12 Main results obtained from flexural pre-cracking tests and creep test by LAB-13

Specimen	f_L (MPa)	$f_{R,p}$ (MPa)	$CMOD_p$ (μm)	$CMOD_{pri}$ (μm)	$CMOD_{pr}$ (μm)	$f_{R,c}$ (MPa)	I_c (%)
S-B1-258/2	3.91	2.18	500.3	423.7	–	1.33	61.05
S-B1-259/2	4.02	3.41	500.2	386.6	–	1.45	42.47

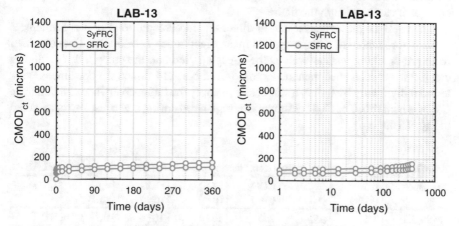

Fig. 7.14 CMOD-time curves obtained in flexural creep test by LAB-13

scatter in the delayed behaviour of the specimens of the same fibre material. The mean creep index I_c applied was 55.60% with a high CV of 12.2%.

The LAB-13 participant tested in flexural creep two SFRC specimens. Since there was only one frame available for the RRT, it was not possible to test any SyFRC specimen. Data obtained from pre-cracking tests as well as the stress applied and the creep index I_c during creep test for each specimen are exposed in Table 7.12. The deformations registered during creep test by LAB-13 are represented in Fig. 7.14. Both specimens were tested in the same frame, with the result of a mean creep index of 51.76% but with a CV of 25.4%. It is important to highlight that this participant suffered an unexpected issue when inducing the load in the frame and it took near 75 min to induce the target load in the creep rig. This fact has a significant influence on the instantaneous deformations and the further creep coefficients and must be considered in the analysis.

7.2 Direct Tension Tests

Direct tension creep tests were performed by 2 laboratories and a total of 10 specimens were tested in creep as shown in Table 4.1. A brief introduction of the results obtained in direct tension tests will be given for each participant.

The LAB-11 participant tested in direct tension creep a total of 6 specimens, 3 of each fibre material, in a multi-specimen set-up with three specimens by frame where a mean creep index I_c of 50.47% was applied with a CV of 7.1%. Data obtained from pre-cracking tests as well as the stress applied and the creep index I_c during creep test for each specimen are exposed in Table 7.13. The COD deformations registered during creep test are represented in Fig. 7.15. The results obtained for both groups of FRCs are consistent and the SyFRC specimens are clearly differentiated from SFRC ones, as it can be seen in the figure.

The LAB-16 participant tested in direct tension creep a total of 4 specimens, 2 of each fibre material, with an applied mean creep index I_c of 50.0%. Data obtained from pre-cracking tests as well as the stress applied and the creep index I_c during creep test for each specimen are exposed in Table 7.14. Since the specimens were tested individually by frame, it was easier to apply the target creep index and thus the CV of is near 0%. The COD deformations registered during creep test are represented in Fig. 7.16.

Table 7.13 Main results obtained from direct tension pre-cracking tests and creep test by LAB-11

Specimen	f_L (MPa)	$f_{R,p}$ (MPa)	COD_p (μm)	COD_{pri} (μm)	COD_{pr} (μm)	$f_{R,c}$ (MPa)	I_c (%)
-MS6_E2	3.61	0.75	200	107	107	0.38	50.71
-MS6_C	3.71	0.77	200	112	112	0.38	49.68
-MS8	3.97	0.73	200	106	106	0.35	47.70
-SF1C	2.65	1.29	200	115	115	0.59	45.94
-SF2C	3.34	1.09	200	118	118	0.58	53.07
-SF4E2	3.24	1.01	200	119	119	0.56	55.75

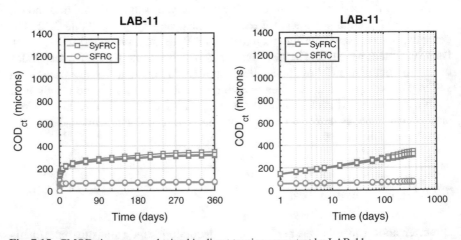

Fig. 7.15 CMOD-time curves obtained in direct tension creep test by LAB-11

Table 7.14 Main results obtained from direct tension pre-cracking tests and creep test by LAB-16

Specimen	f_L (MPa)	$f_{R,p}$ (MPa)	COD_p (μm)	COD_{pri} (μm)	COD_{pr} (μm)	$f_{R,c}$ (MPa)	I_c (%)
M-B2-217	3.21	0.48	141	85	–	0.24	50.00
M-B2-219	3.13	0.81	137	78	–	0.40	50.00
S-B2-398	3.27	1.50	205	129	–	0.75	50.00
S-B2-400	2.91	0.67	190	147	–	0.33	50.00

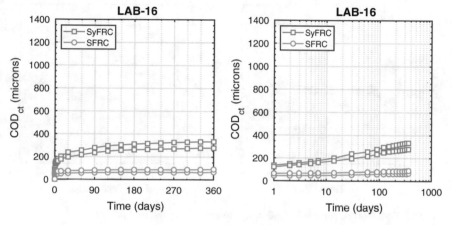

Fig. 7.16 CMOD-time curves obtained in direct tension creep test by LAB-16

The delayed deformations obtained in direct tension creep tests performed by both participant laboratories are quite similar in magnitude and behaviour.

7.3   Square Panel Tests

Square panel creep tests were performed by 3 laboratories and a total of 24 specimens were tested in creep as shown in Table 4.1. A brief introduction of the results obtained in square panel creep tests will be given for each participant.

The LAB-12 participant tested in flexure creep a total of 4 square panel specimens in two creep frames in a double-specimen set-up: one frame for 2 SyFRC and another frame for two SFRC specimens. Data obtained from pre-cracking tests as well as the stress applied and the creep index I_c during creep test for each specimen are exposed in Table 7.15. The mean creep index I_c applied was 64.31% F_L with a CV of 2.2%. The central deflection deformations registered during creep test are represented in Fig. 7.17. The LAB-12 participant performed pre-cracking tests up-to 2 mm deflection and the creep index is referred to F_L. The net load applied in the frame was about 30 kN.

Table 7.15 Main results obtained from pre-cracking and creep test on square panels by LAB-12

Specimen	F_L (kN)	$F_{R,p}$ (kN)	δ_p (μm)	δ_{pri} (μm)	δ_{pr} (μm)	$F_{R,c}$ (kN)	I_c (%)
M-B1-036	46.85	67.63	1936	–	–	30.00	64.03
M-B1-037	48.05	61.62	1954	–	–	30.00	62.44
S-B1-260	45.66	66.49	2064	–	–	30.00	65.71
S-B1-261	46.12	59.45	2015	–	–	30.00	65.05

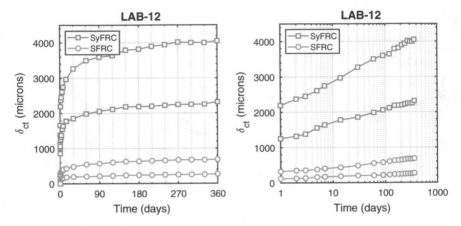

Fig. 7.17 CMOD-time curves obtained in square panel creep test by LAB-12

The LAB-15 participant tested individually in flexure creep a total of 12 square panel specimens: 6 of each fibre material. The pre-cracking tests were performed until the first crack at LOP and the creep index is referred to F_L. Data obtained from pre-cracking tests as well as the stress applied and the creep index I_c during creep test for each specimen are exposed in Table 7.16. The load applied in all the frames was 51.6 kN which correspond to the 120% of the mean F_L obtained from all the pre-cracking tests. This procedure leads to an applied mean creep index I_c of 117.28% F_L and a CV of 14.7%. The central deflection deformations registered during creep test are represented in Fig. 7.18.

Regarding the SyFRC specimens, it can be observed that one specimen (M-B2-144) develops quite higher deformations than the rest of the group, even if the applied creep index was not the highest one. Something similar occurs with the S-B1-266 specimen for SFRC case, what implies that the specimens under the highest creep index do not correspond with those with highest deformations.

The LAB-18 participant tested in flexure creep a total of 8 square panel specimens, 4 of each fibre material, with an applied mean creep index I_c of 49.91% and a CV of 0.1%. Data obtained from pre-cracking tests as well as the stress applied and the creep index I_c during creep test for each specimen are exposed in Table 7.17. The central deflection deformations registered during creep test are represented in Fig. 7.19. As it can be observed, one specimen of each fibre material (M-B2-149 and

Table 7.16 Main results obtained from pre-cracking and creep test on square panels by LAB-15

Specimen	F_L (kN)	$F_{R,p}$ (kN)	δ_p (μm)	δ_{pri} (μm)	δ_{pr} (μm)	$F_{R,c}$ (kN)	I_c (%)
M-B1-040	51.0	–	–	–	–	51.6	101.18
M-B1-041	37.0	–	–	–	–	51.6	139.46
M-B1-042	40.0	–	–	–	–	51.6	129.00
M-B2-142	55.0	–	–	–	–	51.6	93.82
M-B2-143	45.0	–	–	–	–	51.6	114.67
M-B2-144	38.0	–	–	–	–	51.6	135.79
S-B1-264	55.0	–	–	–	–	51.6	93.82
S-B1-265	40.0	–	–	–	–	51.6	129.00
S-B1-266	42.0	–	–	–	–	51.6	122.86
S-B2-349	38.0	–	–	–	–	51.6	135.79
S-B2-350	45.0	–	–	–	–	51.6	114.67
S-B2-351	53.0	–	–	–	–	51.6	97.36

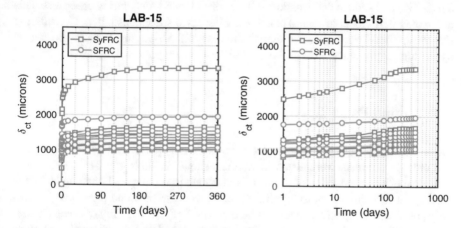

Fig. 7.18 CMOD-time curves obtained in square panel creep test by LAB-15

Table 7.17 Main results obtained from pre-cracking and creep test on square panels by LAB-18

Specimen	F_L (kN)	$F_{R,p}$ (kN)	δ_p (μm)	δ_{pri} (μm)	δ_{pr} (μm)	$F_{R,c}$ (kN)	I_c (%)
M-B2-147	–	66.00	3000	–	–	33.0	49.99
M-B2-148	–	57.50	2970	–	–	28.7	49.84
M-B2-149	–	71.00	3000	–	–	35.4	49.81
M-B2-150	–	73.00	3000	–	–	36.5	49.98
S-B2-354	–	60.00	2970	–	–	29.9	49.90
S-B2-355	–	62.50	3000	–	–	31.2	49.88
S-B2-356	–	66.00	3000	–	–	33.0	49.99
S-B2-357	–	72.00	2970	–	–	35.9	49.90

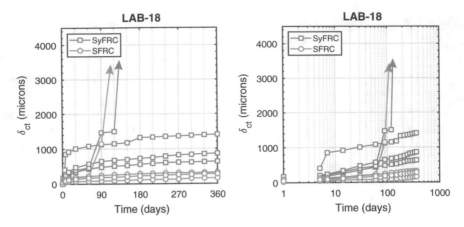

Fig. 7.19 CMOD-time curves obtained in square panel creep test by LAB-18

S-B2-356) had sudden deformation at 90 days and then failed during creep test. This is the only participant in square panel test where two specimens failed. Considering that the applied load was similar than in the case of LAB-12, this fact may be caused by the higher pre-crack level of 3 mm achieved during pre-cracking tests.

7.4 Round Panel Tests

Flexural creep test on round panel specimens were only performed by one laboratory and only 4 specimens were tested in creep as shown in Table 4.1. This section will provide a brief introduction of the results obtained in round panel creep tests by the participant. It is important to highlight that, in the case of round panel tests, it was only possible to conduct creep test for 90 days due to the availability of creep frames.

The LAB-17 participant tested in flexure creep a total of 4 round panel specimens of SyFRC, with an applied mean creep index I_c of 51.9% $F_{R,p}$ (30.8% if referred to F_L) and a CV of 11.6%. Data obtained from pre-cracking tests as well as the stress applied and the creep index I_c during creep test for each specimen are exposed in Table 7.18. The delayed central deflection deformations registered during creep test are represented in Fig. 7.20. Even if round panels were individually tested in

Table 7.18 Main results obtained from pre-cracking tests and creep test on round panels by LAB-17

Specimen	F_L (kN)	$F_{R,p}$ (kN)	δ_p (μm)	δ_{pri} (μm)	δ_{pr} (μm)	$F_{R,c}$ (kN)	I_c (%)
M-B1-043	26.24	15.90	2054	1025	–	9.29	58.43
M-B1-044	27.72	18.44	1977	1009	–	9.09	49.30
M-B2-151	28.25	14.93	1855	1029	–	6.7	44.88
M-B2-152	28.23	16.04	2054	1106	–	8.85	55.17

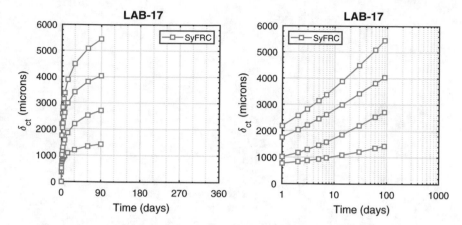

Fig. 7.20 CMOD-time curves obtained in round panel creep test by LAB-17

creep, a significant scatter in terms of instantaneous and delayed deformation can be observed.

Chapter 8
Analysis of the RRT Results

Aitor Llano-Torre, Pedro Serna, Emilio Garcia-Taengua, Rutger Vrijdaghs,
Hans Pauwels, Clementina del Prete, Nicola Buratti, Raúl L. Zerbino,
Wolfgang Kusterle, and E. Stefan Bernard

Abstract The analysis of the results of the round-robin test (RRT) comprised many steps. The variation of the main variables such as the number of samples tested, environmental conditions, pre-crack level or creep index and mechanical performance were analysed among all the RRT participants. In addition, the main methodologies

A. Llano-Torre (✉) · P. Serna
Institute of Concrete Science and Technology ICITECH, Universitat Politècnica de València (UPV), Valencia, Spain
e-mail: aillator@upv.es

P. Serna
e-mail: pserna@cst.upv.es

E. Garcia-Taengua
School of Civil Engineering, University of Leeds, Leeds, UK
e-mail: E.Garcia-Taengua@leeds.ac.uk

R. Vrijdaghs
Building Materials and Building Technology Section, KU Leuven, Louvain, Belgium
e-mail: rutger.vrijdaghs@kuleuven.be

H. Pauwels
NV BEKAERT SA, Zwevegem, Belgium
e-mail: hans.pauwels@bekaert.com

C. del Prete · N. Buratti
Department of Civil, Chemical, Environmental and Materials Engineering DICAM, University of Bologna, Bologna, Italy
e-mail: clementina.delprete2@unibo.it

N. Buratti
e-mail: nicola.buratti@unibo.it

R. L. Zerbino
LEMIT-CIC and Faculty of Engineering UNLP, La Plata, Argentina
e-mail: zerbino@ing.unlp.edu.ar

W. Kusterle
OTH Regensburg, Regensburg University of Applied Sciences, Regensburg, Germany
e-mail: wolfgang@kusterle.net

A. Llano-Torre and P. Serna (eds.), *Round-Robin Test on Creep Behaviour
in Cracked Sections of FRC: Experimental Program, Results and Database Analysis*,
RILEM State-of-the-Art Reports 34, https://doi.org/10.1007/978-3-030-72736-9_8

147

were independently analysed to search significances between the participant procedures. Analysis of the described procedures, test setups and equipment provided relevant information about such existent differences between participants in terms such as load configuration, boundary conditions or frames construction. The creep coefficient and crack opening rate (COR) parameters were defined and analysed. The complementary data number of fibres in the cross section was analysed to assess the influence in the long-term residual performance. Finally, the correlation between different fibre-reinforced concrete (FRC) specimens as well as the influence of ageing of concrete matrix were analysed.

8.1 Analysis of the Main Variables of the RRT

Since this international round-robin test (RRT) programme comprises so many specimens and different procedures, lots of variables were involved and had influence on the RRT results. Therefore, a first approach of main variables and parameters analysis is exposed to check the boundary conditions of the analysis and conclusions of the RRT. The information about the main variables considered by each participant, procedure description and equipment used is widely described in Chap. 5.

8.1.1 Samples Tested on Creep Test by Laboratory and Methodology

A total of 124 fibre-reinforced concrete (FRC) samples were tested in creep in the RRT. The number of specimens tested, as well as the methodology performed depend on the availability of creep frames at each laboratory. Figure 8.1 depicts the number of tested specimens by laboratory and the methodology followed.

Most laboratories performed only one methodology, except for LAB-11 and LAB-12 that could perform two different methodologies: LAB-11 performed flexural creep tests and direct tension test on cored specimens, whereas LAB-12 performed creep tests on flexure for prismatic specimens and square panels. Figure 8.1 shows that the most extended procedure is the Flexure creep tests performed by 12 laboratories, where up to 86 specimens (69% of the specimens tested in creep) were tested following similar criteria but with significant differences regarding the equipment and procedure. Direct tension creep was assessed in 10 specimens by 2 laboratories, which represent 8% of the total specimen tested, as seen in Fig. 8.2. In the case of creep test on panels, the classification was made by the standard associated to the creep test procedure and the shape of specimen (square or round). The square panel

E. Stefan Bernard
TSE Technologies in Structural Engineering Pty Ltd, Sydney, Australia
e-mail: s.bernard@tse.net.au

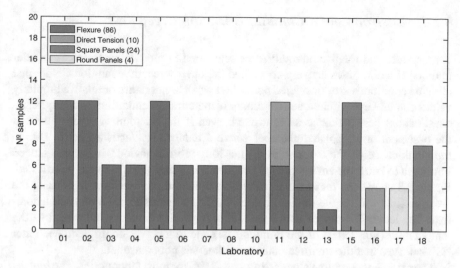

Fig. 8.1 Number of samples tested by laboratory

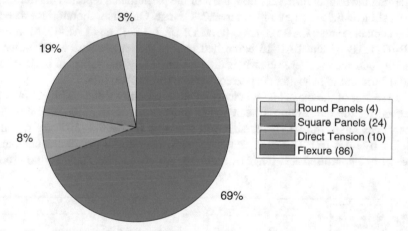

Fig. 8.2 Samples distribution between different creep methodologies performed in the RRT

creep tests were performed in 24 specimens tested by 3 laboratories (19% of total specimens), whereas only one laboratory performed creep test on 4 round panels (3% of the total specimens).

8.1.2 Environmental Conditions During Creep Test

Temperature and relative humidity were registered in most laboratories. A smaller number of laboratories actively controlled the environmental conditions. Note that climatic conditions are considered controlled when temperature or relative humidity values can be kept within a certain range. If the climatic conditions are only registered cannot be considered as controlled, even if the location of creep frames is the basement or an isolated chamber where a reduced scatter is expected. Due to the significance of these variables on the long-term behaviour, it became indeed important to know the environmental conditions in which the specimens were tested. Figure 8.3 shows the mean values of temperature and relative humidity during the creep test, as well as the minimum and maximum values reached on each laboratory. The letter "Y" placed above the x-axis for each laboratory indicates that the temperature (first letter) or the humidity (second letter) was controlled. The letter "N" indicates that the environmental condition was not controlled.

The mean temperature values are around 20 °C for most laboratories. The highest mean values over 24 °C are registered in LAB-05, LAB-06, LAB-13 and LAB-16. Regarding the temperature variation, most of the participants reported that temperature was hold in the average temperature ±2 °C range. Only five laboratories exceeded this acceptable range: LAB-03, LAB-06, LAB-07, LAB-15 and LAB-17. Moreover, LAB-03, LAB-07 and LAB-15 controlled neither the temperature nor the relative humidity and such scatter is clearly justified. Conversely, the values of LAB-06 and LAB-17 present high scatter despite controlling the temperature.

The scatter of the mean and boundary relative humidity values is bigger than the observed for the temperature (mean values range from 35 to 65% and the boundaries values range from 15% registered in Japan by LAB-08 to 80% registered in Belgium by LAB-07). This such scatter between laboratories makes evident that an agreed or standardised test range is required. An acceptable range for controlled

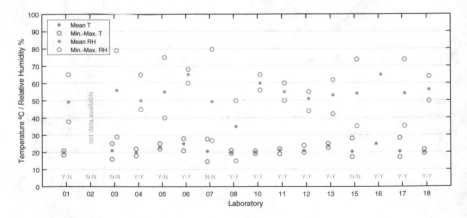

Fig. 8.3 Temperature and relative humidity registered during creep tests

relative humidity can be defined around ±10% or even closer. Only seven laboratories (LAB-04, LAB-06, LAB-10, LAB-11, LAB-12, LAB-13 and LAB-18) remain in this range, all of which control the relative humidity. LAB-08 and LAB-17 show variations of respectively 35% and 38.5% despite reporting that the relative humidity was controlled. Amongst the rest of the laboratories that do not control the relative humidity, the highest variation was registered by LAB-07 (52.8%) and the lowest variation was reported by LAB-01 (27.4%). Even though LAB-01 did not control the relative humidity, the creep frames were located inside an isolated chamber not significantly affected by the outside climatic conditions, as seen in Fig. 7.2.

8.1.3 Pre-crack Level Reached on Pre-cracking Tests

Although the target pre-crack level was agreed by the participants and the TC, the procedure adopted by each lab might induce variations in some parameters such as the pre-crack levels achieved. Figure 8.4 depicts the mean pre-crack value of each laboratory arranged by creep test procedure and the reference measure is reflected in the legend for each methodology.

The pre-crack levels achieved by all the participants conducting flexural creep tests in beams are close to the target value (CMOD of 500 μm, as described in Sect. 6.2.1). Note that LAB-03 and LAB-07 controlled pre-cracking by measuring the beam deflection (δ_p), which was converted into $CMOD_p$ values. A slightly bigger variation around the target pre-cracking level was achieved in the direct tension test

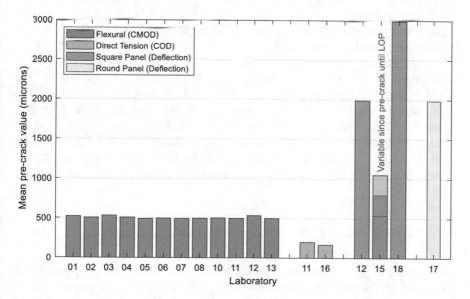

Fig. 8.4 Mean pre-crack values reached on pre-cracking tests

performed by LAB-11 and LAB-16 (COD of 200 μm). This may be attributed to the difficulty in the control of the crack opening in the direct tension pre-cracking process. In the case of laboratories performing panel creep test, LAB-12 and LAB-17 achieved the target deflection of 2 mm. By contrast, LAB-18 achieved 3 mm deflection and LAB-15 experienced a variable deflection as the pre-cracking stage was interrupted upon reaching LOP. The difference in procedure adopted by LAB-18 and LAB-15 is likely to affect the comparison of creep test results with the other two laboratories undertaking panel creep tests.

8.1.4 Mechanical Properties: Residual Strengths in Flexure

The FRCs designed for the RRT were characterized and classified at different ages alongside the creep test execution. Specimens tested in flexure creep can be also analysed and classified depending on their residual strengths obtained in pre-cracking flexure tests. Since most participants performed the pre-cracking test following EN 14651 procedure using a three-point bending test (3PBT) setup, the Limit of Proportionality (LOP) f_L and the residual strength $f_{R,1}$ can be directly calculated. In cases of laboratories using a four-point bending test (4PBT) in the pre-cracking stage, stress values were converted into equivalent 3PBT results following the equations presented in Sect. 8.3.1. Figure 8.5 shows the histograms of f_L and $f_{R,1}$ measured in the beam pre-cracking and the stress applied during the creep stage ($f_{R,c}$). Note that results approximate a normal distribution. $f_{R,c}$ average is 1.47 MPa with values ranging from 0.8 to 2.2 MPa. This range fits the criterion of the reference value for the service limit state (SLS) cracking analysis, as defined in Sect. 6.2.2. The findings derived from the RRT are conditioned to the ranges of variables studied. Any extrapolation should be avoided.

The f_L and $f_{R,1}$ for each prismatic specimen in the pre-cracking stage are summarised in Fig. 8.6. The mean f_L is 4.47 MPa with a range from 3.4 to 5.8 MPa. Since the fibres do not directly affect the LOP, similar f_L are obtained for both macro-synthetic fibre-reinforced concrete (SyFRC) and steel fibre-reinforced concrete (SFRC). LAB-05 obtained the highest residual strength values at LOP for both fibre materials. $f_{R,1}$ shows higher scatter, as normally observed in FRC. The mean $f_{R,1}$ is 2.91 MPa with values ranging from 1.8 to 4.2 MPa. A higher scatter is observed for SFRC due to the lower number of fibres crossing the cracked section.

8.1.5 Creep Index I_c

As explained in Sect. 6.2.4, the target creep index (I_c) for flexural creep and direct tension creep test was defined as 50% of $f_{R,p}$ at CMOD of 0.5 mm and COD of 0.2 mm, respectively. For both square and round panel creep tests I_c was set to 60% of f_L. Figure 8.7 shows I_c for all specimens tested classified in methodologies. The

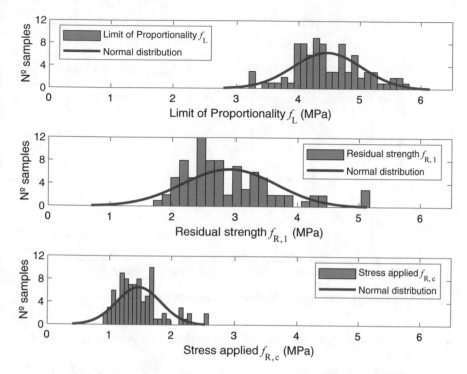

Fig. 8.5 Histogram of residual strength data distribution for flexural creep test specimens

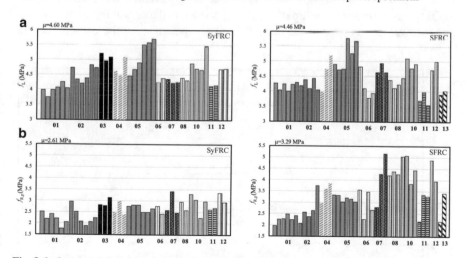

Fig. 8.6 f_L (**a**) and $f_{R,p}$ (**b**) values for the specimens tested by the different laboratories

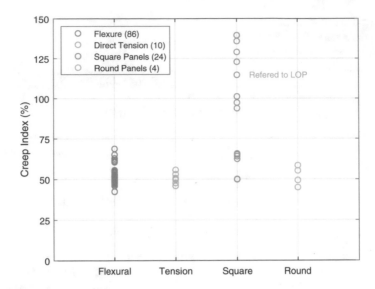

Fig. 8.7 Creep index I_c applied values for different methodologies

number of specimens tested according to each methodology is presented in brackets in the legend.

The average I_c and the obtained scatter by each laboratory is presented in Fig. 8.8 by means of box and whisker plot. As a general observation, the scatter of I_c for multi-specimen creep test setup is usually higher due to the impossibility of selecting one load value that satisfies the desired I_c of all specimens in the column. On the

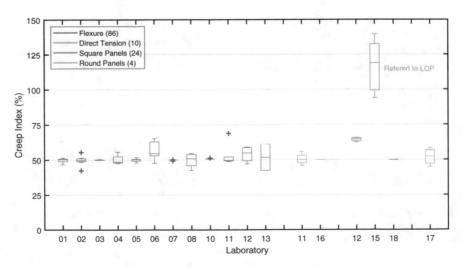

Fig. 8.8 Applied creep index I_c by participant for all methodologies

contrary, a smaller scatter was observed in the I_c adopted in single specimen setups at LAB-03, LAB-07, LAB-10 or, LAB-16. The use of multi-specimen setup is more extended in methodologies like flexure and direct tension whereas, both panel creep tests are usually performed in a single specimen setup, saving some exceptions such as LAB-12.

Different I_c were applied in the case of square panel creep test due to the specific procedures adopted in each laboratory. LAB-12 applied on each frame 60% of the F_L of the specimens. Despite using multi-specimen frames with two specimens tested simultaneously, the creep index approximates the target value. The applied creep index by LAB-18 on individual creep frames was 50% of the $F_{R,p}$ obtained at 3 mm deflection in the pre-cracking tests, as seen in Fig. 8.8. The creep index scatter depicted by LAB-15 may be explained by the application of 120% of the mean F_L obtained in three specimens tested in flexure. Since LAB-15 applied the same reference load to all specimens, the creep index became scattered due to the inherent dispersion of the FRC and fibre distribution along cracks. In the case of round panel test, the applied creep index by LAB-17 was 50% of the $F_{R,p}$ at 2 mm deflection and mean value around 51.9%. Despite using single specimen setup, a significant scatter is observed due to forklift issues as explained in Sect. 5.15.4.

The I_c applied in the direct tension creep test approximate the target value with only small scatter due to the multi-specimen setup for LAB-11. LAB-16 applied 50% of the $f_{R,p}$ using a setup with a single specimen.

Considering the big number of specimens in flexural tests, I_c was also analysed for SyFRC and SFRC separately, as depicted in Fig. 8.9. SyFRC specimens show more homogeneous I_c, whereas higher scatter can be observed for SFRC specimens possibly due to the lower number of steel fibres bridging the crack. Thus, it can be concluded that in case of low dosages of fibres, the flexure creep tests procedure may lead on relatively high scattered results and therefore a minimum dosage or minimum structural performance should be recommended for this procedure.

Figure 8.10 shows the I_c histogram plot for flexure test procedure. Most specimens of flexural creep (87.2%) had I_c ranging from 45 to 55%. The out-of-range creep index values only represents 12.8% of the total specimens tested in flexural creep. It should be noted again that the conclusions obtained from the analysis of the results of this RRT are conditioned to this studied range of creep index and extrapolation is not recommended.

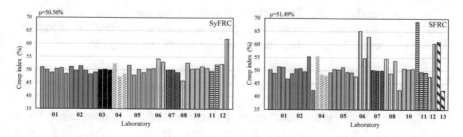

Fig. 8.9 Creep index values for the specimens tested by the different laboratories

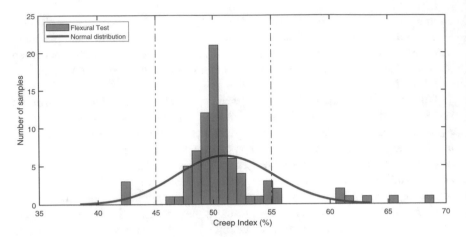

Fig. 8.10 Histogram of creep index data distribution in flexure creep tests

8.2 Analysis of Different Methodologies

This section presents a general analysis of interdependence among test parameters considered in the RRT and their influence on the evolution of the delayed displacements. The different methodologies were analysed separately.

8.2.1 Flexural Creep Test

This subsection focuses on the analysis of the results of the flexural creep tests on prismatic specimens obtained from the RRT database [18]. The analysis used the first version of the RRT results, as presented in Chap. 7 and Appendix C. The objective was to obtain a preliminary model for the CMOD as a function of time (t) and variables such as creep index (I_c), stress level ($f_{R,c}$), fibre material, and laboratory. The objective was to identify the main influencing variables, determine whether a consistent model can be applied with an acceptable fitting and detect the laboratory dependency.

The mathematical rationale for the definition of the model and its format was explained in relation to the characteristics of the data and the technicalities of multiple regression analysis. In particular, the following aspects were justified:

- Response parameter to be taken as log(CMOD), which was preferable to other options such as the untransformed total $CMOD_{ct}$ values (instantaneous and delayed deformations) or creep coefficients.
- Differences between laboratories to be represented by a categorical (non-quantitative) variable which grouped the laboratories into 'clusters'. The reason

behind this grouping of laboratories into clusters was justified on the grounds that it is easier to handle a variable that takes no more than 4–5 values.

8.2.1.1 Re-evaluating the Grouping of Laboratories

A more systematic approach to the grouping of laboratories into 'clusters' has been followed this time. It consisted of the following stages:

- In consistency with the format of the previous model for CMOD, a preliminary regression analysis was performed to fit a model that relates log(CMOD) to I_c, $f_{R,c}$, t1/2, and fibre material and all possible pairwise interactions between them. That is, all variables except the laboratory.
- The residuals from the above (difference between measured CMOD value and the value predicted by the fitted model, for all observations in the database) were stored in a new variable and squared: Res^2. Since the squared residuals of any regression model contain the scatter not explained by the variables in the model, and the differences between laboratories had not been considered, the squared residuals contained the part of variability in log(CMOD) values that was attributable to differences between laboratories.
- Res^2 values were plotted against the categorical variable 'Laboratory', and statistically significant differences were identified. Any subset of laboratories for which the average Res^2 values were not statistically significantly different were assumed to be in the same cluster.

As a result, it was concluded that three of the laboratories were significantly dissimilar between them and different from all others, which were grouped in two clusters. In total, there were again 5 clusters, where the Cluster A comprises the 70% of the specimens tested in flexure as seen in Fig. 8.11. Upon examination of

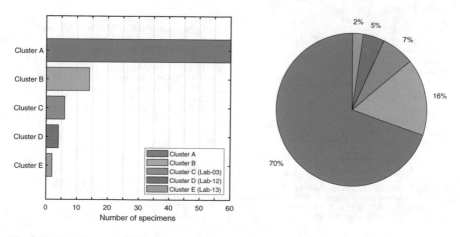

Fig. 8.11 Significance of different clusters in specimens tested in flexure creep

the procedural similarities and differences between the different laboratories, some possible reasons were identified:

- Cluster 'A': laboratories 01, 02, 04, 05, 06, 08, and 11. In all these, the pre-cracking setup consisted in a 3PBT scheme, and the creep test was set up as a 4PBT scheme.
- Cluster 'B': laboratories 07, and 10. In these, both the pre-cracking and creep setups consisted in a 3PBT scheme.
- Lab-03: Pre-cracking and creep setups were 4PBT, and specimens were unnotched.
- Lab-12: Both pre-cracking and creep setups were 4PBT.
- Lab-13: Pre-cracking and creep setups were 4PBT, and the initial time t_{ci} significantly displaced with respect to the other laboratories.

8.2.1.2 Regression Analysis: Model for CMOD

The multiple regression analysis was performed following different model formats and the best performance in terms of both goodness of fit and diagnostics (normality of the residuals, homoscedasticity, etc.) was achieved considering the log-transformed time. This was coded as $\log(time + 0.001)$, as otherwise $\log(time)$ was not defined for $time = 0$.

The final model and corresponding significance tests are summarized in Fig. 8.12. The R-squared was 0.796, i.e. 79.6%, which is very good considering not only the intrinsic variability of creep measurements and of the properties of cracked FRC specimens but also the fact that the data comes from different sources, using different instruments and operated by different people, etc.

```
lm(formula = log(CMOD) ~ LabCluster + Ic + I(Ic^2) + fRc + I(fRc^2) +
as.factor(Material) + log(time + 0.001) + as.factor(Material):log(time + 0.001))

Coefficients:
                                     Estimate  Std. Error t value Pr(>|t|)
(Intercept)                         5.6545824   0.8080152   6.998 3.85e-12 ***
LabClusterB                         0.3048121   0.0248313  12.275  < 2e-16 ***
LabClusterLab03                    -0.9629530   0.0320816 -30.016  < 2e-16 ***
LabClusterLab12                     0.6169645   0.0365892  16.862  < 2e-16 ***
LabClusterLab13                    -0.8199260   0.0530513 -15.455  < 2e-16 ***
Ic                                 -0.0656011   0.0289743  -2.264  0.02370 *
(Ic^2)                              0.0007954   0.0002638   3.015  0.00261 **
fRc                                 0.8489773   0.1584032   5.360 9.60e-08 ***
(fRc^2)                            -0.2003312   0.0470442  -4.258 2.18e-05 ***
(Material)Synthetic                 0.5972796   0.0229621  26.011  < 2e-16 ***
log(time + 0.001)                   0.0461613   0.0030804  14.985  < 2e-16 ***
(Material)Synthetic*log(time + 0.001)  0.0830007   0.0044959  18.462  < 2e-16 ***
---
Signif. codes:  0 '***' 0.001 '**' 0.01 '*' 0.05 '.' 0.1 ' ' 1


Residual standard error: 0.317 on 1553 degrees of freedom
  (199 observations deleted due to missingness)
Multiple R-squared:  0.7955, Adjusted R-squared:  0.794
F-statistic: 549.1 on 11 and 1553 DF,  p-value: < 2.2e-16
```

Fig. 8.12 Final model and corresponding significance for CMOD in flexure creep tests

Table 8.1 Coefficients K_0, K_1 and K_2 to be applied in the equation for CMOD

Depending item	Options	K_0	K_1	K_2
Material	SFRC	5.6546		0.04616
	SyFRC	6.2518		0.12916
Laboratory/Cluster	Cluster A		±0.000	
	Cluster B		+0.305	
	Cluster C (Lab-03)		−0.963	
	Cluster D (Lab-12)		+0.617	
	Cluster E (Lab-13)		−0.820	

Directly from the previous table and considering that $\log(t + 0.001) \cong \log(t)$, the equation for CMOD is the following:

$$\log(CMOD) = K_0 + K_1 + K_2 \log(t) - 0.0656\, I_c + 0.00079\, I_c^2$$
$$+ 0.849\, f_{Rc} - 0.2\, f_{Rc}^2 \tag{8.1}$$

where the coefficients K_0 and K_2 depend on the fibre materials and K_1 depends on the laboratory. Their values are defined in Table 8.1:

8.2.1.3 Final Equation for CMOD

Exponentiating both sides of the equation above in order to remove logarithms, and after some manipulation, the model can be rewritten as follows:

$$CMOD = K \cdot F(I_c) \cdot F(f_{Rc}) \cdot F(t) \tag{8.2}$$

where:

- The 'basic' CMOD curve is $F(t)$, is a function of time t (days), and the fibre material:

$$F(t) = \begin{cases} 285.6\, t^{0.04616} & (steel) \\ 518.9\, t^{0.1292} & (synthetic) \end{cases}$$

- The multiplying function $F(f_{R,c})$ accounts for the effect of the stress applied (in MPa):

$$F(f_{Rc}) = \exp\left(0.849\, f_{Rc} - 0.2\, f_{RC}^2\right)$$

- The multiplying function $F(I_c)$ accounts for the effect of the creep index (in %):

$$F(I_c) = \exp\left(0.00079\, I_c^2 - 0.0656\, I_c\right)$$

- The constant K is a multiplier that depends on the laboratory:

$$K = \begin{cases} 1.00 & (A) \\ 1.36 & (B) \\ 0.38 & (Lab03) \\ 1.85 & (Lab12) \\ 0.44 & (Lab13) \end{cases}$$

This model proves that the only factor that intervenes directly in the relationship between CMOD and time is the fibres material, represented by what has been called the 'basic' curve. Moreover, the effect of the differences introduced by the testing procedure, and changes in the creep index or the stress applied can be decoupled into univariate functions that are 'multipliers' of the basic curve. In the following sections, some plots are presented to describe the main characteristics of each of the three functions above.

8.2.1.4 Basic Curve F(T)

The basic curve plots for SyFRC and SFRC mixes are shown in Fig. 8.13. This curve represents only the effect of the time. The actual CMOD is evaluated by multiplying the F(t) value by factors accounting for the cluster, $f_{R,c}$ and I_c. For instance, for cluster A and average values of I_c and $f_{R,c}$, the multiplying factor is 0.62.

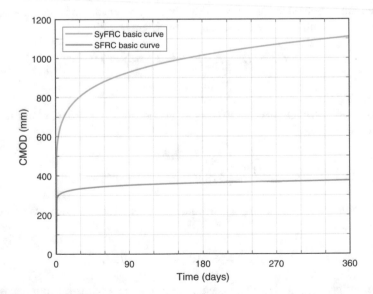

Fig. 8.13 Basic curve F(t) for the two FRC mixes considered

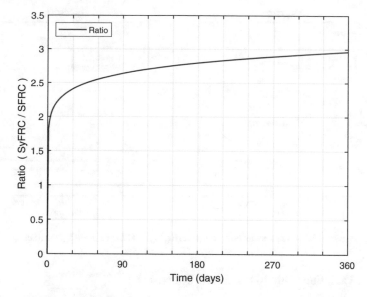

Fig. 8.14 CMOD ratio between SyFRC and SFRC mixes, all other parameters being equal

The ratio between F(t) assuming SyFRC and SFRC mixes indicates how many times CMOD values are higher for SyFRC than for SFRC while other factors are equal. This gives:

$$\frac{F(t)_{SyFRC}}{F(t)_{SFRC}} = \frac{518.9\,t^{0.1292}}{285.6\,t^{0.0462}} - 1.817\,t^{0.083}$$

Note that the ratio function increases with time. The SyFRC mix gave CMOD values that are between 2.2 and 2.8 times those obtained by SFRC, as shown in Fig. 8.14.

8.2.1.5 Variation Between Laboratories

Figure 8.15 shows the effect of applying the different values of the multiplier K on the basic curve for SyFRC and SFRC mixes (line in bold corresponds to Cluster A with K = 1).

The maximum difference between laboratories is approximately of 4 times: $(1.85 - 0.38)/0.38 = 3.87$. CMOD values measured by any can be up to 85% higher or up to 62% lower than the average, basic curve.

It is also important to emphasize that the relative magnitude of these variations is independent of the material of the fibres, but the absolute difference is not, as seen in Fig. 8.15.

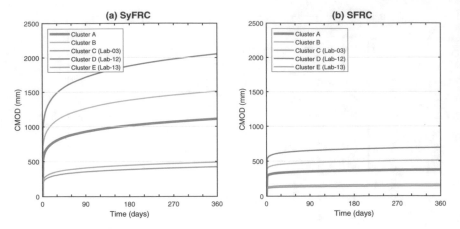

Fig. 8.15 Between-laboratories variation considering SyFRC (**a**) or SFRC (**b**) mixes

8.2.1.6 Variations Due to the Creep Index: F(I$_c$)

The multiplier related to the creep index shown in Fig. 8.16a increases dramatically for high creep index values. However, this is an example of the risk of extrapolation outside the area of validity, and it must be approached with caution as the information in the database does not cover all values. This exponential trend suggests a potential error if results reported here are extrapolated. As seen in Fig. 8.10, the multiplying function of the *creep index* is most representative of I_c between 40 and 60%. Therefore, Fig. 8.16b provides a more accurate representation of the multiplying effect of I_c on CMOD values.

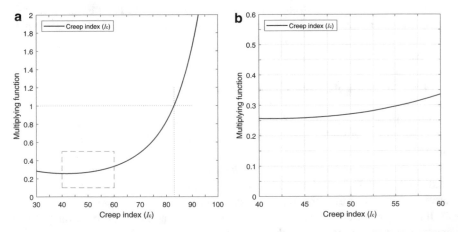

Fig. 8.16 Values of the multiplying function F(I_c) versus *creep index* values (**a**), multiplying function F(I_c) for I_c between 40 and 60% (**b**)

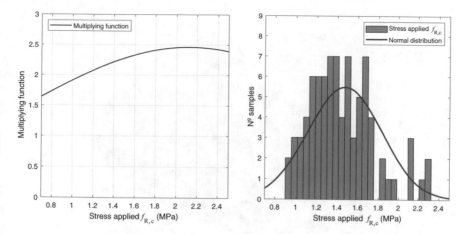

Fig. 8.17 Values of the multiplying function F($f_{R,c}$) for $f_{R,c}$ values between 0.8 and 2.4 MPa (**a**); Histogram of the stress applied values $f_{R,c}$ in the database (**b**)

The mathematical definition of this multiplying function aids the understanding of the basic function F(t). Since this multiplier takes a value of 1 for a *creep index* of 83%, it can be said that the basic function F(t) and therefore the CMOD curves in Figs. 8.13 and 8.15 correspond to the case $I_c = 83\%$. However, the information in the database, which is what this model has been based upon, does not include cases where I_c is 83%, as can be seen in the histogram in Fig. 8.5. Therefore, it is probably better to consider that the basic function F(t) and therefore the CMOD curves in Figs. 8.13 and 8.15 constitute *an upper limit* to the CMOD curves.

8.2.1.7   Variations Due to the Stress Applied: F($F_{R,C}$)

The multiplier that depends on the level of stress applied (plotted in Fig. 8.17a) increases steadily and stabilises at a value of approximately 2.5 for stress levels above 1.9 MPa. Similarly to the observed for the load factor, the analysis and discussion of the effects of the stress level on CMOD values as per the plot in Fig. 8.17 must be limited to the range between 0.8 and 2.3 MPa approximately, in correspondence with the limitations of the information in the database as per the histogram in Fig. 8.17b.

8.2.1.8   Representational Possibilities of the Model

The model developed supports the analysis and interpretation of the effect of different factors affecting the evolution of creep. For example, the combined effect that variations in the creep index and the stress level have on CMOD values can be analysed in detail if the product of the two multipliers $F(I_c) \, F(f_{R,c})$ is plotted against I_c and $f_{R,c}$ values, as shown in Fig. 8.18. The effect of any of the parameters considered

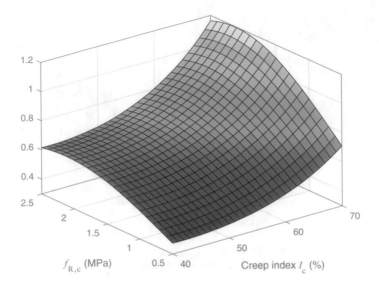

Fig. 8.18 Combined effect of the two multiplying functions: creep index and stress applied

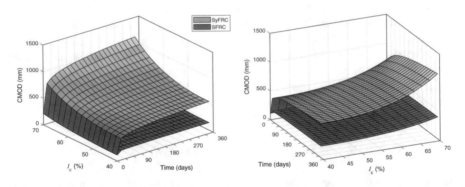

Fig. 8.19 CMOD(t) for SyFRC and SFRC mixes, assuming lab in cluster A and $f_{R,c} = 1.5$ MPa

can also be explored in a more direct manner if the basic curve of CMOD vs time is extended, for example, to cover a range of different creep index. In Fig. 8.19, the model is plotted against creep index and time considering either SyFRC mix (in green) or SFRC mix (in purple).

8.2.1.9 Concluding Remarks

In this first approach, a preliminary version of the RRT database [18] was analysed and some minor errors were detected among the provided data. After revision and update of the data, a clear dependency on the procedural and methodological aspects adopted by each laboratory was found. Since most of laboratories performed similar procedures, most specimens can be grouped in the same cluster (cluster A). This indicates that the definition of a standardised procedure can be feasible and would contribute to the repeatability and the comparison of creep test results obtained in different laboratories. Although this study should be applied again for the final version of the database, it could not be developed now since more data of other creep parameters were required.

Not included in this analysis are several other aspects that are also relevant, such as the effects on variability and the post-processing of the model equations in order to obtain characteristic values or to obtain robust creep coefficients values that could be included in structural evaluations. A further study of these additional aspects should consider maximum and minimum creep values with a reasonable confidence interval.

8.2.2 Direct Tension Creep Test

In the RRT of the RILEM TC 261-CCF, 10 specimens were tested in uniaxial tension at 2 laboratories (5 SyFRC specimens and 5 SFRC specimens). LAB-11 tested 6 cylindrical specimens, 3 with each fibre type. The diameter of the notched section is 78 mm on average, resulting in an average cracked surface area of 4860 mm^2. LAB-16 tested 4 prismatic specimens in tension with a square notched section equation to an average surface area of 6700 mm^2 (+38%). The small number of specimens limits the statistical analysis. As in the analysis of the flexural tests, an attempt is made to highlight the most important factors governing the creep behaviour, such as fibre type, *creep index* (I_c) or residual strength ($f_{R,1}$).

8.2.2.1 Global Behaviour

In the following figures (Fig. 8.20), the raw data results are shown for SyFRC and SFRC mixes, both in terms of the total deformation in μm and in terms of the creep coefficient.

SyFRC samples show bigger creep than those of SFRC, both in terms of absolute crack width (COD) as well as creep coefficient. After one year under sustained load, the deformation is still increasing albeit at a reduced pace. Power laws, in the form of $x = a \cdot t^b + c$, can be fitted to the experimental data. Considering that at t = 0, COD = point E and φ = 0, the following power laws can accurately ($r^2 > 0.99$ for all curves) fit the creep data:

$$COD_{steel} = 6.184 \cdot t^{0.2345} + 51.8$$
$$COD_{synt} = 95.18 \cdot t^{0.1776} + 53.2$$
$$\varphi_{steel} = 0.1188 \cdot t^{0.2383}$$

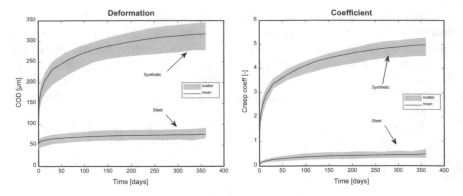

Fig. 8.20 Delayed deformations and creep coefficients obtained from direct tension creep tests

$$\varphi_{synthetic} = 1.792 \cdot t^{0.1777} \tag{8.3}$$

In the present form, both the crack width and creep coefficient tend to infinity (also assumed in the flexural tests). Figure 8.21 shows the ratio between the average COD or φ of the SFRC and SyFRC. On average, the average crack opening is 4 times larger in SyFRC than for SFRC, while the creep coefficient is roughly 10 times bigger. Both ratios roughly stabilize after 3 months under sustained load. Note that the COD ratios in uniaxial tension (around 4) are considerably bigger than those estimated in Sect. 8.2.1.4 for specimens subjected to flexure creep. This is expected as the uniaxial tensile test is a more direct test of the fibre material rather than the FRC matrix. Additionally, the stresses in the compressive zone in flexural creep tests can (partly) offset the intrinsic fibre differences, leading to lower ratios in bending.

Fig. 8.21 Ratio evolution of the average COD and the creep coefficient φ between SyFRC and SFRC mixes

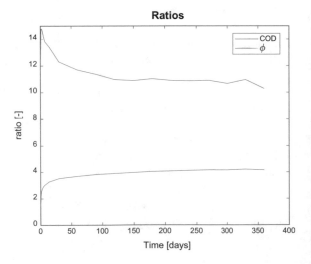

8.2.2.2 *Influence of* Creep Index (I_c)

LAB-16 reported that effective I_c for all the specimens was equal to 50%, so no analysis can be conducted on their experiments, which leaves only 6 specimens for the evaluation of the influence of I_c. The recorded effective I_c were:

- SFRC: 45.9%, 53.1%, and 55.8%.
- SyFRC: 50.7%, 49.7%, and 47.7%.

The analysis focusses on the presence of outliers and their possible influence on the results identified through Grubbs' test. The maximum calculated z-score is 1.1 for 45.9% (ST), indicating that this value is 1.1 times the standard deviation from the sample average. Given the low number of samples, this value is below the critical z-score (1.2 at 5% significance level) and we can assume that no significant difference is observed between the I_c. Since no statistically significant difference is found, the low number of samples does not allow identifying the influence of the I_c.

8.2.2.3 *Influence of* $F_{R,1}$

The analysis may be performed for the residual strength, which is recorded for all 10 samples:

- SFRC: 1.29, 1.09, 1.01, 1.50, and 0.67 MPa.
- SyFRC: 0.75, 0.77, 0.73, 0.48, and 0.81 MPa.

Only one outlier is identified through the Grubbs' test: the 4th SyFRC specimen (0.48 MPa) has a z-score of 1.74 which is above the critical limit of 1.72. Therefore, the results of this specimen are further investigated, both in terms of COD as φ. If the creep deformations of this result are significantly different from the others, then the influence of $f_{R,1}$ needs to be further investigated. However, the creep results of the outlier specimen are in line with the others and no statistically significant difference is observed.

8.2.2.4   Influence of the Laboratory

For each type of fibre, only 5 specimens are available: 3 from LAB-11 and 2 from LAB-16. Despite the low number of tested specimens in direct tension, no significant differences between laboratories were found and results for the two groups seem to show a good repeatability.

8.2.2.5   Influence of the Fibre Type

Finally, the influence of the fibre type is discussed. From the analysis of the residual strength, it is clear that a very different $f_{R,1}$ is obtained. Through a double-sided student-t hypothesis test, it is found that the mean residual strength is indeed significantly different: $P(T \leq t) = 0.04$. Furthermore, the global creep results indicate a large difference between both fibre types. The fibre type seems indeed the determining factor in the behaviour under uniaxial creep.

8.2.2.6   Conclusion

Similar to the analysis of the flexural tests, the direct tension creep results can be described and fitted with a power function:

$$COD(t) = K \cdot f_1(LR) \cdot f_2(f_{R1}) \cdot f_3(t, fiber\ type) \tag{8.4}$$

The limited number of specimens does not support the identification of statistically significant influence of the *creep index*, residual strength, or the laboratory, which were assumed as 1 in the formulation:

$$K = f_1(LR) = f_2(f_{R1}) = 1 \tag{8.5}$$

If compared with the flexure creep test, the direct tension creep test is a more complex and unstable testing procedure and is less common in research and practice and far to be standardized. Since the number of participants is reduced and the procedure control is higher, the study of the influence of parameters like creep index, concrete strength or testing procedure was not possible. Therefore, based on the available data, it must be concluded that the creep behaviour shows a clear dependence on the fibre type. An expression similar of this obtained for flexural test is found in direct tension showing a similar tendency on time but, in direct tension creep test, the influence of the fibre type is higher than that observed for the flexural test.

8.2.3   Square and Round Panel Creep Test

Three of the participating laboratories performed tests on FRC square panels. Each laboratory used the same basic panel size according to EN 14488-5 [28] with significant differences in the applied methodology of the creep tests. In addition, one participant performed creep test on FRC round panels in compliance with ASTM C1550 [29]. Table 8.2 provides an overview of such differences among participants.

The most notable differences are in the pre-crack level and the applied load level in the creep stage. The pre-crack level is the point to which the panels were loaded in a standard EN 14488-5 [28] frame before they were moved to the creep frames.

The pre-crack level is an important variable as it is closely related to the load applied during the creep test. The limit of proportionality (LOP) in a panel subjected to EN 14488-5 [28] test occurs at a deflection lower than 1 mm. After the LOP is reached, the panels still show considerable hardening behaviour up to 4–5 mm

Table 8.2 Differences between test methodologies applied in the participating laboratories

Methodology	Square panel			Round panel
Laboratory	LAB-12	LAB-15	LAB-18	LAB-17
Specimens/frame	2	1	1	1
Pre-crack level	2 mm deflection	LOP	3 mm deflection	2 mm deflection
Load application	Hydraulic	Lever arm	Lever arm	Dead load
Load calibration	Pressure gage	Load cell	Load cell	Load cell
Load control	Daily check, weekly adjust	Not controlled	Not controlled	Electronic logger
Creep index	60% F_L	120% mean F_L	50% $F_{R,p}$	50% $F_{R,p}$
Applied load (kN)	30	54	32	8.45
Temperature	22 ± 2 °C	N/A	N/A	N/A
R. humidity	55 ± 5%	N/A	N/A	N/A

deflection. This means that in the case of the different methodologies chosen by the three laboratories, most panels will still show some hardening behaviour in the creep stage.

8.2.3.1 Total Deformation During Creep Stage

Looking at the evolution of the total creep deformation is not statistically relevant as the dispersion amongst the individual test results is far too high. Figure 8.22 shows that the total creep deformation for SFRC panels is between 150 and 2000 μm after a year. The total deformation after a year for SyFRC panels ranges from 700 and 4000 μm.

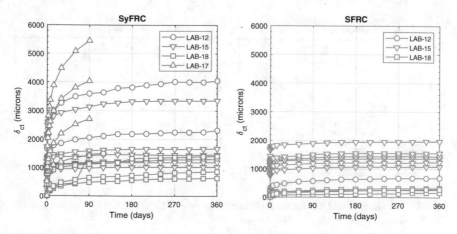

Fig. 8.22 Total deformation (δ_{ct}) during creep stage on the square panels

8.2.3.2 Pre-crack Deformation

Each laboratory aimed for a different total deformation in the pre-cracking stage: LAB-12 and LAB-17 aimed at an initial (real) deflection of 2 mm, LAB-18 stopped the pre-cracking stage at 3 mm and LAB-15 stopped pre-cracking at the limit of proportionality (LOP), which corresponds to a real deformation between 536 μm and 1 mm and average δ_p of 798 μm.

Considering the creep coefficients at 180 days, test results show that a higher initial deformation leads to higher creep coefficient during the creep stage. As mentioned before, this can be attributed to the fact that the panels have less capacity left for any hardening behaviour when further deformation occurs. The rate of increment decreases considerably over time. The creep coefficient for a given pre-cracking deformation (Point D) is higher in SyFRC than in SFRC. This difference is most noticeable when the specimen had higher pre-crack deformation. The laboratory with the lowest pre-crack level (Point D < 1000 μm) applied the highest load during the creep stage (51.6 kN). The other laboratories applied a similar load (30–33 kN) during the creep stage. Note that round panels were not included in this analysis since they were only tested until 90 days.

The number of available data points is limited to n = 11 for SFRC as well as for SyFRC panels. The results for SyFRC panels show more scatter, specially at higher pre-cracking deformations. A non-linear (exponential) regression proved to be the best fit for the available data. Figure 8.23a shows the proposed fit for SFRC panels. This model corresponds to the given data with a relatively narrow 95% prediction interval and a standard error of 0.198. Figure 8.23b shows the model for SyFRC panels. The higher amount of scatter in the experimental data yields a model that is less accurate with a standard error of 2.117.

Figure 8.24 shows a summary of both models as well as the creep coefficient ratio ($\varphi^{180}_{w,c\,synthetic}/\varphi^{180}_{w,c\,steel}$) based on these models. This ratio shows a linear relation to the pre-cracking deflection. For a pre-cracking deflection of 500 μm, the creep

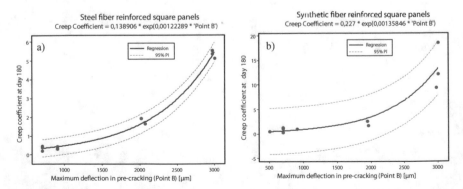

Fig. 8.23 Non-linear fitted line for the creep coefficient of SFRC (**a**) and SyFRC (**b**) panels at day 180 of the creep stage

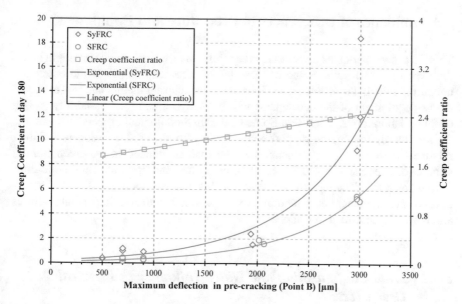

Fig. 8.24 Summary of the regression analysis on the creep coefficient

coefficient of SyFRC is 1.75 bigger than for SFRC. At a pre-cracking deflection of 3 mm, this ratio increases to 2.5.

8.2.3.3   Recommendations for Future Testing

The total creep deformation is not an adequate parameter to judge the creep performance of a material. To circumvent the inherent scatter in these measurements, the analysis should focus on relative parameters such as the creep coefficient $\varphi_{w,c}^{j}$, defined as the ratio between the initial deformation in the creep stage and the deformation at a certain time j. After a test duration of $j = 180$ days, the initial quick increase of the creep coefficient is gradually slowing down to a stable pace. Any differences between (fibre) materials should have clearly emerged at that stage.

Another factor for consideration is the pre-cracking extent. Results show that higher pre-cracking displacements lead to an increase in the differences between fibre types. The applied load during the creep stage is less important for the outcome of a test.

Note that EN 14488-5 [28] and the creep test use a hyperstatic setup. Isostatic tests like three- or four-point bending tests are useful to derive material properties. A hyperstatic test shows how a material performs in a specific system. The recommendations listed above are useful to find differences between materials within a feasible timeframe. It is important to assess these parameters considering a specific application.

8.3 Influence of Differences in Procedure and Equipment

There are many variables involved in creep flexural tests performed that might affect the experimental results. Some of these factors are controllable, while others are uncontrollable. In the following sections, the most relevant factors in terms of procedure or equipment of the participants will be analysed and their variability among different laboratories, discussed. Some of the methodological aspects analysed are the load configuration, the support boundary conditions, used transducers and their relative location in the specimens, how the load is induced in the creep frames, number of specimens tested in the same creep test and environmental conditions during creep test. The information used for this analysis about the different procedures and equipment of each laboratory was provided by the participant laboratories and is widely described in Chap. 5.

8.3.1 Load Configuration and Notch Influence in Flexural Creep Tests

Since each laboratory developed their methodology adapted to their national standards, there are multiple load configurations used in the RRT for both flexural pre-crack and creep tests. It is common among the participants to develop their methodology starting from the EN 14651 standard with a 3PBT configuration in notched specimens for pre-cracking tests (8 of 12 laboratories). Most laboratories (10 out of 12) switch to a 4PBT configuration during the creep tests, which enables a more stable stacking of specimens in a column. Figure 8.25 presents both loading configurations representing the most extended procedures.

Table 8.3 shows the load configuration used in pre-cracking and creep tests, as well as if specimens were notched or not. As it can be observed, this analysis has grouped in clusters laboratories with similar methodology and configuration in line with the classification discussed in Sect. 8.2.1.1 based on the creep test results. All participants of cluster A except LAB-08 follow the same procedure with 3PBT in pre-cracking test and 4PBT for creep test in notched specimens. Note that LAB-06 is the only one from cluster A that performs a 4PBT type 4P-3 configuration defined

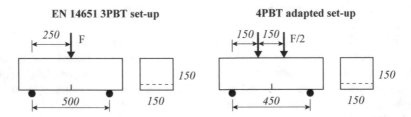

Fig. 8.25 3PBT flexural load standard configuration EN 14651 [22] versus adapted 4PBT set-up

Table 8.3 Flexural configurations used in pre-cracking and creep tests classified by cluster

Methodology	Cluster	Laboratory	Notch	Load configuration		Type
				Pre-crack	Creep	
Flexural	A	LAB-01	Yes	3PBT	4PBT	4P-1
		LAB-02	Yes	3PBT	4PBT	4P-1
		LAB-04	Yes	3PBT	4PBT	4P-1
		LAB-05	Yes	3PBT	4PBT	4P-1
		LAB-06	Yes	3PBT	4PBT	4P-3
		LAB-08	Yes	4PBT	4PBT	4P-1
		LAB-11	Yes	3PBT	4PBT	4P-1
	B	LAB-07	Yes	3PBT	3PBT	3P-1
		LAB-10	Yes	3PBT	3PBT	3P-1
	C	LAB-03	No	4PBT	4PBT	4P-2
	D	LAB-12	Yes	4PBT	4PBT	4P-1
	E	LAB-13	Yes	4PBT	4PBT	4P-4

in Table 8.4. The cluster B is composed by LAB-07 and LAB-10, which performed creep tests on individual specimens using a 3PBT. Three laboratories perform only 4PBT in all tests, but each participant uses a different configuration leading to their classification in clusters C, D and E. There is only one participant (LAB-03) that following the German and Austrian guidelines [24, 25] performed both tests on unnotched specimens in a 4PBT configuration (cluster C).

Figure 8.26 shows the three different combinations of configurations found in the database. Most laboratories (50%) adopted 3PBT for pre-cracking and 4PBT for creep tests, 33% adopted 4PBT for both tests, while 17% adopted 3PBT only.

It is important to highlight that in the case of different load configuration between pre-cracking and creep tests, a formulation must be used to convert the load that leads to the same stress level. In the case of 3PBT configuration, load and stress are related by the EN 14651 equation:

$$M = \frac{F}{2} \cdot \frac{L}{2} = \frac{FL}{4} \quad \text{where} \quad f_{R,j} = \frac{6M_j}{bh_{sp}^2} = \frac{3F_jL}{2bh_{sp}^2} \tag{8.6}$$

whereas for those laboratories that use a 4PBT adapted configuration:

$$M = \frac{F}{2} \cdot \frac{L}{3} = \frac{FL}{6} \quad \text{where} \quad f_{R,j} = \frac{6M_j}{bh_{sp}^2} = \frac{F_jL}{bh_{sp}^2} \tag{8.7}$$

Despite extended use of 4PBT adapted set-up for flexural creep, there exist up to four different configurations performed in the RRT and even also differences in specimen dimensions due to limitations in creep frame size. Different configurations in terms of distance between supports is found among laboratories using 4PBT during

Table 8.4 Flexural load configurations performed during creep tests in the RRT

Type	Creep flexural configuration	Participants	Cluster	Stress/load ratio ($f_{R,c}/F_{R,c}$) Theoretical	Real
3P-1	250 / 500 / 150 / 150	LAB-07 LAB-10	B B	1:3.13	1:3.12 1:3.13
4P-1	150 150 / 450 / 150 / 150	LAB-01 LAB-02 LAB-04 LAB-05 LAB-08 LAB-11 LAB-12	A A A A A A D	1:5.21	1:5.36 1:5.39 1:5.03 1:5.48 Not available 1:5.68 1:5.20
4P-2	150 150 / 450 / 150 / 150	LAB-03	C	1:7.5	1:7.49
4P-3	175 150 / 500 / 150 / 150	LAB-06	A	1:4.46	1:4.47
4P-4	200 200 / 600 / 150 / 75	LAB-13	E	1:1.95	1:1.95

Fig. 8.26 Static configurations for pre-cracking and creep tests found in the RRT database

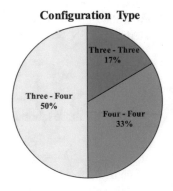

Configuration Type

Three - Three 17%

Three - Four 50%

Four - Four 33%

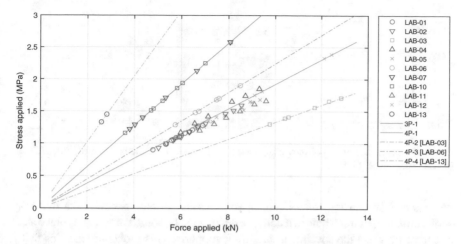

Fig. 8.27 Stress/load ratio for the different specimens tested in flexural creep

the creep test, which are summarised in Table 8.4. The theoretical stress/load ratios
($f_{R,c}/F_{R,c}$) calculated according to Eqs. (8.6) and (8.7) for the five load configurations
were obtained and compared with those ratios obtained from the database results.

This Stress/Load ratio comparison is also depicted in Fig. 8.27 and served as
double check of the RRT results. Standard 3PBT and 4PBT load configuration (9
participants) are represented by continuous lines whereas the three dash lines represent the adapted 4PBT flexural configuration (3 participants). The most extended
configuration for flexural creep test is 4P−1 as observed in Table 8.4.

8.3.2   Multi-specimen Configuration

The use of a multi-specimen configuration allows increasing the number of specimens
tested by frame at the same time. Despite the common use of the multi-specimen
configuration, there are multiple configurations available for multi-specimen setup
in the RRT depending on the construction of the creep frame or the relative positions of the specimens in the creep frame. This section provides an analysis of the
multi-specimen configurations available and the possible influence on the long-term
behaviour.

One of the main differences among multi-specimen configurations is the influence
of the individual deformations of one specimen on the stability of the rest of the
specimens. If one specimen of the frame suffers a collapse during the creep test,
this sudden deformation may cause instability in the system, reduction of load in the
frame or even interrupt the creep test of the other specimens in the column. In order
to avoid this influence, some laboratories place steel load plates between specimens
that provide a certain independency to the specimens. These steel load plates carry

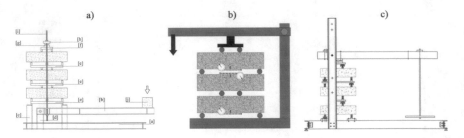

Fig. 8.28 Different multi-specimen setups used in flexural creep methodologies: (**a**) LAB-01, (**b**) LAB-05 and (**c**) LAB-11

on load and supporting rollers and, in some cases, an emergency brake that may stop a broken specimen before it collapses. Conversely, some laboratories place steel cylinders between the specimens as rolling supports. This procedure implies that a sudden displacement of any specimen will affect to the rest of the specimens of the frame and their results. Moreover, this procedure usually intermediate specimens to be turned 180° in order to ensure stability in the column.

The three different multi-specimen setups used in the RRT are depicted in Fig. 8.28: (a) multiple specimens with intermediate steel plates without specimen inversion (LAB-01), (b) multiple specimens without intermediate plates and inverted specimen (LAB-05) and (c) multiple specimens with intermediate steel plates and inverted specimen (LAB-11).

In single specimen creep test setups, higher accuracy in the creep index can be achieved than in the multiple-specimen setup. A summary of multi-specimen setup, the creep index average values as well as the coefficient of variation of each participant is exposed in Table 8.5.

The lowest coefficient of variations corresponds to those laboratories that perform creep test with single specimens (LAB-07, LAB-10, LAB-03, LAB-16 and LAB-18). Notwithstanding, there are participants testing individual specimens that obtain a quite high coefficient of variation (LAB-15 and LAB-17). Such a variation in LAB-15 can be explained due to the different criteria of the reference load applied, since they applied the same load to all frames (50% of the mean f_L obtained from the characterization test of the batches). In general, due to the difficulty of the creep procedure and the long duration of the creep tests, it is highly recommended to build creep frames where each specimen in the column can move separately without influence the load or displacements of the rest of the specimens. The use of creep frames where the local collapse of one specimen does not affect or case the interruption of the creep test of the rest of the specimens is recommended.

Table 8.5 Differences between participants regarding the multi-specimen setup and influence on the creep index variation

Methodology	Cluster	Laboratory	Multi	N° Sp.	Inverted	Independent	Mean I_c	CV (%)
Flexural	A	LAB-01	Yes	3	No	Yes	49.91	2.7
		LAB-02	Yes	3	Yes	No	49.79	5.9
		LAB-04	Yes	3	No	Yes	50.09	6.4
		LAB-05	Yes	3	Yes	No	50.02	2.1
		LAB-06	Yes	3	No	Yes	56.34	11.6
		LAB-08	Yes	2	Yes	No	49.75	9.7
		LAB-11	Yes	3	Yes	Yes	53.02	14.7
	B	LAB-07	No	1	No	Yes	49.90	0.7
		LAB-10	No	1	No	Yes	50.70	0.5
	C	LAB-03	No	1	No	Yes	50.06	0.5
	D	LAB-12	Yes	2	Yes	Yes	55.60	12.2
	E	LAB-13	Yes	2	Yes	No	51.76	25.4
Direct tension		LAB-11	Yes	3	–	No	50.47	7.1
		LAB-16	No	1	–	Yes	50.00	0.0
Square panel		LAB-12	Yes	2	Yes	No	64.31	2.2
		LAB-15	No	1	No	Yes	117.28	14.7
		LAB-18	No	1	No	Yes	49.91	0.1
Round panel		LAB-17	No	1	No	Yes	51.94	11.6

8.3.3 Support Boundary Conditions

The quality construction of the support rollers is quite relevant to assure the desired support boundary conditions. Given the variety on support rollers construction by participants, a classification of both support and load rollers is proposed depending on the degrees of freedom. This consensus on the roller classification will avoid misunderstandings during boundary conditions definitions and results interpretation. Considering a general support, the most relevant movements (degrees of freedom) that could affect the behaviour of the specimens are indicated in Fig. 8.29. The proposed classification evaluates three different degrees of freedom: (I) rotation in X axis, (II) rotation in Z axis and (III) translation in X axis. Note that the X axis corresponds to the longitudinal axis of the beams.

The different degrees of freedom are indicated with a three-character code composed by two different symbols: X or O. The former (X) represents a blocked movement while the latter (O) represents a free movement. Following this definition, all support or load rollers used in this RRT were classified as summarised in Table 8.6. The support classification, description and any picture of the construction is given for all cases available.

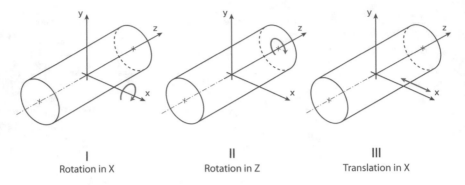

I	**II**	**III**
Rotation in X	Rotation in Z	Translation in X

Fig. 8.29 Classification of the degrees of freedom of supports

Table 8.6 Boundary conditions of creep frame supports by participant

Code	Description	Support construction
XXX		
XOX		
OXX		
OOX		
XOO		

Table 8.7 Boundary conditions of creep frame supports by participant

Methodology	Cluster	Laboratory	Support rollers		Loading points		Multi	N°
			A	B	A	B		
Flexural	A	LAB-01	OXX	XOX	OXX	XOX	Yes	3
		LAB-02	XXX	XXX	XXX	XXX	Yes	3
		LAB-04	OXX	XOX	OXX	XOX	Yes	3
		LAB-05	OXX	XOX	OXX	XXX	Yes	3
		LAB-06	OOX	OOX	OOX	OOX	Yes	3
		LAB-08	XOO	XOO	XXX	XXX	Yes	2
		LAB-11	OOX	XOX	OOX	XOX	Yes	3
	B	LAB-07	XOO	XOO	XXX	–	No	1
		LAB-10	XOO	XOO	XOO	–	No	1
	C	LAB-03	XXX	XXX	XOO	XXX	No	1
	D	LAB-12	OOX	XOX	OOX	XOX	Yes	2
	E	LAB-13	XOX	XOX	XOX	OXX	Yes	2
Direct tension		LAB-11	OOO	–	OOO	–	Yes	3
		LAB-16	OOO	–	OOO	–	No	1
Square panel		LAB-12	XXX	–	XXX	–	Yes	2
		LAB-15	XXX	–	XXX	–	No	1
		LAB-18	XXX	–	XXX	–	No	1
Round panel		LAB-17	XOX	–	XOX	–	No	1

The classification of all the supports and load boundary conditions of all the participants of the RRT are represented in Table 8.7 and reflects the high variability regarding the boundary conditions of the creep test procedures in the different laboratories. Moreover, the multi-specimen configuration information aids to check how varies the support and load roller freedom for different configurations.

In the case of both square and round panel test, since the flexural creep test configuration is the same as in EN 14488-5 [28] and ASTM C1550 [29], the boundary conditions among the different participants are the same. Something similar occurs in direct tension creep test where the free movements and rotation of the supports is always ensured to avoid bending forces in tension tests.

Unfortunately, in the case of flexural creep test of prismatic specimens, there are multiple combinations of the different rollers support construction. The support boundary conditions show more variations in the case of multiple-specimen setup where steel plate of specific rollers must be designed. Laboratories that perform single specimen setup creep tests for both 3PBT (LAB-07 and LAB-10) or 4PBT (LAB-03) have the same support classification XOO, where solid steel rollers are free to rotate in axis Y and translate in axis X. On the contrary, this type of solid steel rollers located between two specimens became a support classified as XXX since both the rotation and the translation are blocked due to the simultaneous deformation

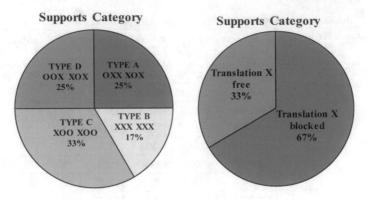

Fig. 8.30 Types of beam supports found among the RRT participants

of both specimens. The most extended classification of the support rollers among the participants that perform multiple-specimen creep test is one XOX for the first support and either OXX or OOX for the second support. The translation in axis X of the supports is restricted in all cases, excepting for the LAB-08. Four main combinations of support types were identified among the RRT participants for flexural creep tests, as highlighted by Fig. 8.30.

8.3.4 Load Application System in Frame

Regarding the way of the application of the load in the creep frames, there are four different procedures available depending on the construction of the frames. The most extended (12 out of 18 laboratories) is by means of a lever arm where the load of the counterweights is multiplied by a certain factor depending on the length of the lever arm. The hydraulic jack is only used in four laboratories. LAB-08 uses screwed bars calibrated with strain gauges whereas LAB-18 uses a dead load system directly over the specimens. A classification of participants by methodology and load application is given in Table 8.8. Most laboratories using lever arm are from cluster A. On the contrary, all laboratories that used hydraulic actuators were identified by the analysis as different clusters and thus, significantly different. By means of this analysis can be concluded that with a lever arm system, a similar methodology is detected, and more repetitive results may be obtained.

8.3.5 Time in Which Load Is Applied

This parameter plays a significant role in the creep coefficient calculation and therefore must be well defined for further experimental programmes. The time in which

Table 8.8 Classification of participants by time in which load is applied in the creep frame

Methodology	Load system	Cluster	Laboratory	Lever arm position
Flexural	Lever arm	A	LAB-01	Lower
			LAB-02	Lower
			LAB-04	Lower
			LAB-05	Upper
			LAB-06	Lower
			LAB-11	Upper
		B	LAB-07	Upper
		C	LAB-03	Lower
	Hydraulic jack	B	LAB-10	–
		D	LAB-12	–
		E	LAB-13	–
	Screw bars	A	LAB-08	–
Direct tension	Lever arm	–	LAB-11	Upper
		–	LAB-16	Upper
Square panel	Hydraulic jack	–	LAB-12	–
	Lever arm	–	LAB-15	Upper
		–	LAB-18	Upper
Round panel	Dead load	–	LAB-17	–

the load is applied in the creep frame has direct influence on the instantaneous deformations, since most of deformation occurs when the load is applied. The longer the time of load application, the higher the instantaneous deformation $CMOD_{ci}$ registered. Considering that in the creep coefficient calculation the instantaneous deformation $CMOD_{ci}$ is in the denominator of Eq. (8.8) defined in Sect. 8.4, the time in which load is applied in the creep frame must be controlled and restricted.

The time in which the desired load is applied depends mainly on variables like the load application system or if a multiple-specimen setup is adopted. There is a discussion about if the stacking time should be considered in the t_{ci} or if it should be discarded due to the very low load level during stacking time. The consideration of the stacking time in the t_{ci} may cause an increment on the time in which load is applied but not necessarily an increase of the instantaneous deformations since the load of the upper specimens is usually negligible compared to the load applied during the creep test. The load system, load calibration and the construction of the creep frames are also significant since these items also influence in the loading procedure and time.

The different t_{ci} values are depicted for all participants in Fig. 8.31. There is a high scatter among time adopted to apply the load in different laboratories, with values ranging from 2 (LAB-02) to 4475 (LAB-13) seconds.

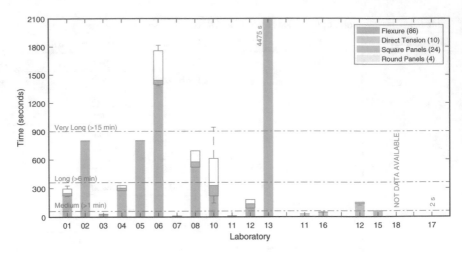

Fig. 8.31 Time in which the load was applied in the creep frames by the participant

Due to such a scatter, a first tentative classification by t_{ci} for the different participant laboratories was proposed in four different levels as follows:

- Short: $t_{ci} < 60$ s (1 min)
- Medium: 60 s $< t_{ci} < 360$ s (6 min)
- Long: 360 s $< t_{ci} < 900$ s (15 min)
- Very Long: $t_{ci} > 900$ s (15 min).

The thresholds of the different t_{ci} are also depicted in Fig. 8.31 by means of dashed red lines. Following this classification, the laboratories were ordered by time in which load was applied in an ascendant order in Table 8.9. The table also include the load system information as well as if a multi-specimen setup is adopted.

The influence of the time in which load is applied t_{ci} on the instantaneous and short-term $CMOD_{ci}$, $CMOD_{ci}^{10'}$ and $CMOD_{ci}^{30'}$ deformations is illustrated in Fig. 8.32. The instantaneous deformation measured differs significantly when different times t_{ci} are considered: the influence of the t_{ci} on the instantaneous deformations increases for longer durations of load applying. The $CMOD_{ci}$ increases from short to medium values of t_{ci} a 13% whereas from short to very long time increases a 31%. This $CMOD_{ci}$ variation directly affects the creep coefficients and therefore it becomes very important to define the loading process and the $CMOD_{ci}$ parameter in further creep standardised procedures.

When analysing the influence on the different $CMOD_{ci}$ references, a $CMOD_{ci}^{30'}$ variation of 28% was observed for short t_{ci} times (>60 s). Considering the long and very long t_{ci} times, the variation was reduced to 14% and 8% respectively as expected. The longer time considered as t_{ci}, the higher deformations due to sustained load are considered as instantaneous deformations and therefore, the variation between different $CMOD_{ci}$ gets closer. Once again it is concluded how important becomes

Table 8.9 Classification of participants and clusters depending on t_{ci}

Methodology	Classification	Mean t_{ci}	CV (%)	Participant	Cluster	Multi-specimen	Load system
Flexural	Short	10	0.0	LAB-11	A	Yes	Lever arm
		10	0.0	LAB-07	B	No	Lever arm
		30	0.0	LAB-03	C	No	Lever arm
	Medium	135	38.5	LAB-12	D	Yes	Hydraulic jack
		258	17.1	LAB-01	A	Yes	Lever arm
		302	9.8	LAB-04	A	Yes	Lever arm
	Long	425	64.5	LAB-10	B	No	Hydraulic jack
		596	13.2	LAB-08	A	Yes	Screw bars
		800	0.0	LAB-02	A	Yes	Lever arm
		800	0.0	LAB-05	A	Yes	Lever arm
	Very long	1538	12.6	LAB-06	A	Yes	Lever arm
		4475	0.0	LAB-13	E	Yes	Hydraulic jack
Direct tension	Short	30	0.0	LAB-11		Yes	Lever arm
		44.8	15.0	LAB-16		No	Lever arm
Square panel	Short	60	0.0	LAB-15		No	Lever arm
	Medium	142.5	10.5	LAB-12		Yes	Hydraulic jack
		–	–	LAB-18		No	Lever arm
Round panel	Short	2.0	0.0	LAB-17		No	Dead load

to define the loading procedure and the $CMOD_{ci}$ parameter to correctly consider the instantaneous deformations.

8.3.6 Environmental Conditions During Creep Test

In the absence of reference recommendation, each participant performed the creep test where possible in their facilities. This explains the wide range of environmental conditions found in the RRT. A first classification can be done regarding the control or not of the environmental conditions during creep tests. The percentages of laboratories controlling temperature and humidity are reported in Fig. 8.33.

A classification of the participant laboratories by location of the creep test and the control of the climatic conditions is given in Table 8.10. There are two main locations available for the creep test: inside a climatic room or in laboratory ambient

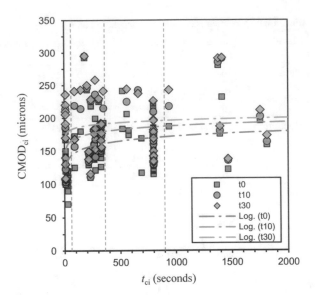

Fig. 8.32 Influence of the time in which load is applied t_{ci} on the instantaneous and short-term $CMOD_{ci}$, $CMOD_{ci}^{10'}$ and $CMOD_{ci}^{30'}$ deformations

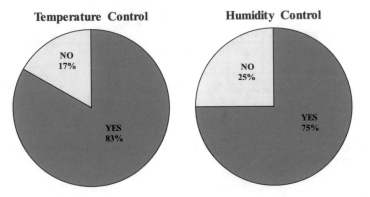

Fig. 8.33 Percentages of laboratories controlling temperature and relative humidity

conditions. In addition, one participant (LAB-03) performs the creep test in the laboratory basement where the air flow and humidity are restricted but not controlled. On the contrary, this participant wrapped the FRC specimens in aluminium sheets to avoid basic shrinkage.

The highest variations in temperature and humidity were registered in laboratories that performed the long-term test in ambient conditions without any control. Even in cases in which temperature or humidity are reported as controlled (i.e. LAB-06 in temperature or LAB-08 in relative humidity) a significant scatter was reported in the environmental conditions.

Table 8.10 Classification of participants by location of the creep test and control of environmental conditions

Methodology	Location	Laboratory	Cluster	Temperature (°C)				Relative humidity (%)			
				Control	Mean	Min	Max	Control	Mean	Min	Max
Flexural	Climatic room	LAB-01	A	Yes	19.8	18.5	20.8	No	49.2	37.7	65.0
		LAB-04		Yes	20.0	18.0	22.0	Yes	50.0	45.0	65.0
		LAB-05		Yes	23.0	22.0	25.0	No	55.0	40.0	75.0
		LAB-06		Yes	25.0	21.0	28.0	Yes	65.0	60.0	68.0
		LAB-08		Yes	20.0	19.4	21.1	Yes	35.0	15.0	50.0
		LAB-11		Yes	21.0	19.0	22.0	Yes	55.0	50.0	60.0
		LAB-10	B	Yes	20.0	19.2	20.8	Yes	60.0	56.0	65.0
		LAB-12	D	Yes	21.0	20.0	24.0	Yes	51.0	44.0	55.0
		LAB-13	E	Yes	23.9	22.8	25.2	Yes	53.1	42.3	61.9
	Laboratory	LAB-02	A	No	–	–	–	No	–	–	–
		LAB-07	B	No	20.6	14.6	27.7	No	49.4	26.8	79.6
	Basement	LAB-03	C	No	21.0	16.0	25.0	No	56.0	29.0	79.0
Direct tension	Climatic room	LAB-11	–	Yes	21.0	19.0	22.0	Yes	55.0	50.0	60.0
		LAB-16	–	Yes	25.0	–	–	Yes	65.0	–	–
Square panel	Climatic room	LAB-12	–	Yes	21.0	20.0	24.0	Yes	51.0	44.0	55.0
	Laboratory	LAB-15	–	No	20.5	17.2	28.4	No	54.1	35.3	73.8
		LAB-18	–	Yes	20.3	19.5	21.6	Yes	56.3	49.9	64.0
Round panel	Climatic room	LAB-17	–	Yes	20.5	17.2	28.4	Yes	54.1	35.3	73.8

The scatter in the temperature is in general smaller than the scatter in relative humidity and thus, it can be concluded that humidity seems to be more difficult to control for the participants. Since the relative humidity influences both creep and shrinkage deformations, a significant improvement on its control is required. The use of a well-insulated testing room is a good starting point towards reducing humidity fluctuations. However, the insulation alone is not enough to ensure a limited variation of the humidity, which requires the adoption of active control measures. If creep tests are performed in well-insulated rooms without any hygrometric equipment or temperature control, the climatic conditions must be described as *"restricted"*. To ensure *"controlled"* climatic conditions, hygrometric and temperature equipment should keep the environmental conditions within the defined ranges.

8.4   Analysis of Creep Coefficient Parameter

The present section analyses the results of the flexural tests on beams in terms of creep coefficient. This parameter provides an indication of the long-term deformation at a time t relative to the initial elastic deformation at time t_0. In the RRT framework, this parameter may be defined either in terms of CMOD or deflection (δ) according to Eqs. (8.8) and (8.9), respectively:

$$\varphi_{w,c}^{j} = \left(CMOD_{ct}^{j} - CMOD_{ci} \right) / CMOD_{ci} \tag{8.8}$$

$$\varphi_{w,c}^{j} = \left(\delta_{ct}^{j} - \delta_{ci} \right) / \delta_{ci} \tag{8.9}$$

The elastic deformation at the denominator of Eqs. (8.8) and (8.9) is defined as the relative deformation from point D to point E of the plot in Fig. 6.2, also known as creep coefficient relative to the creep stage. For that reason, t_0 will be assumed as equal to t_E. Although the creep coefficient referred to origin (considering the residual deformation at the end of pre-cracking tests, O-D in Fig. 6.1) is not taken into account in the following sections, it is reported in the RRT database [18] and in Appendix D of this report.

Figures 8.34 and 8.35 show the mean creep coefficients related to the creep stage computed combining data from all SyFRC and SFRC tested specimens, respectively. Two main differences are observed between these figures: creep coefficient values for SyFRCs is (i) higher and (ii) more scattered than for SFRCs. The first difference depends on the physical behaviour of the two composite materials induced by the fibres, while the second difference might be due to the definition of the elastic deformation, i.e. the definition of t_0, as discussed in the next section.

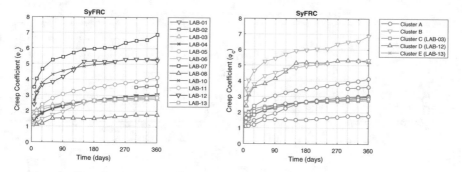

Fig. 8.34 Creep coefficient versus time for SyFRC specimens tested in flexure and classified by cluster

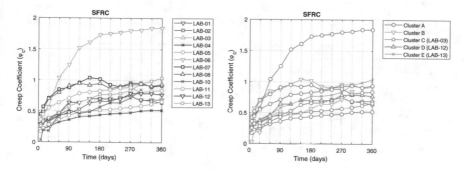

Fig. 8.35 Creep coefficient versus time for SFRC specimens tested in flexure and classified by cluster

8.4.1 Influence of Instantaneous and Short-Term Deformation Definition

There is a crucial discussion about when starts creep or when is stabilised the threshold between instantaneous and delayed deformations. The correct definition of when instantaneous deformation ceases and creep deformation takes over is central for the calculation of the creep coefficient related to the creep stage. Different definitions of t_0 were considered to assess how the concept of initial deformation value affects the creep coefficient. The first approach considers that instantaneous deformation ends immediately after the target applied stress is achieved ($t_0 = t_E$). Moreover, two additional short-term approaches were assessed: $t_0 = t_E + 10'$ and $t_0 = t_E + 30'$ after achieving the target stress.

Figure 8.36 shows values of the creep coefficient at 360 days ($\varphi_{w,c}^{360}$) for each of the aforementioned considerations. For the SyFRC specimens, the maximum mean $\varphi_{w,c}^{360}$ is 6.89 for $t_0 = t_E$, 5.51 for $t_0 = t_E + 10$ min, and 4.41 for $t_0 = t_E + 30$ min. The minimum values of $\varphi_{w,c}^{360}$ decreases from 1.76 to 1.69 when considering $t_0 = t_E + 30$ min rather than $t_0 = t_E$. Considering the SFRC specimens, $\varphi_{w,c}^{360}$ values are less

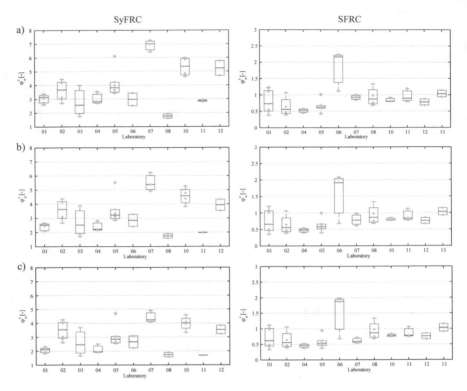

Fig. 8.36 Creep coefficient for both SyFRC and SFRC beams at $t = 360$ days, considering different t_0 times: **a** $t_0 = t_E$, **b** $t_0 = t_E + 10'$ and **c** $t_0 = t_E + 30'$

sensitive to the definition of the elastic deformation. The maximum mean value of this parameter is 1.84 for $t_0 = t_E$, 1.56 for $t_0 = t_E + 10$ min and 1.51 for $t_0 = t_E + 30$ min. The higher sensitivity of the creep coefficient for SyFRC may be attributed to the higher creep rate experienced by the macro-synthetic fibres at early stages. This suggests that a clear and consistent definition of t_0 among laboratories is required for a proper analysis of results, as explained in Sect. 8.4.2.

The definition of the instantaneous deformation is also related to the duration of the load application stage at the very beginning of creep tests, i.e. the time needed to go from point D to point E in the plot of Fig. 6.2. The loading time reported by the laboratories that performed flexural creep tests is plotted in Fig. 8.37 and is clearly very variable. As explained in Sect. 8.3.5, four main groups of laboratories were identified: (i) loading time shorter that 60 s; (ii) loading time between 60 and 360 s; (iii) loading time between 360 and 900 s, and (iv) loading time longer than 900 s. The high variability of the reported values of duration of the loading stage might either indicate an inconsistent definition of the parameter by the different laboratories or significant differences in the experimental procedures adopted to apply loads to the specimens. SyFRC specimens are more sensitive to this parameter and, therefore,

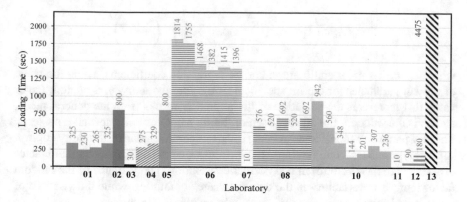

Fig. 8.37 Duration of the load application stage at the beginning of creep tests in different laboratories

part of the variability observed in the creep coefficient values is due to the variability of the loading time.

8.4.2 Statistical Analysis of the Effect of Test Conditions and Mechanical Parameters

In order to understand which parameters might influence the experimental results in terms of creep coefficient, a statistical analysis of the dataset based on mixed effect nonlinear regression models was performed. The main idea of the analysis is to define a mathematical relationship among the different factors listed and classified in Sect. 8.3 in order to identify which factors have statistically significant effect in the creep coefficient and the test results by different laboratories.

Note that the variability of the mechanical properties, some of which cannot be controlled, is in some cases not optimal for statistical analysis. For instance, only a limited number of laboratories reported $f_{R,1}$ values larger than 4.0 MPa for SFRC specimens. Furthermore, these laboratories feature few to no specimens with $f_{R,1}$ less than 3.0 MPa. Therefore, in statistical analysis it might not be possible to infer whether the differences in the results of these laboratories are due to the high $f_{R,1}$ values or to other, possibly unknown, factors. A sound statistical analysis of the effects of the significance of the various parameters, which was not the aim of the RRT, would require a specific design of experiments, aimed at identifying the effect of each parameter and interactions among parameters.

Based on the qualitative and quantitative factors discussed in the previous section, a mathematical relationship that describes the creep coefficient evolution over time for the ith specimen of the jth laboratory was defined as:

$$\varphi_{ij}(t) = \left(\delta_j + c_o + \sum_{k=1}^{N} c_k x_{kij} + \varepsilon_{ij}\right) \cdot \left(\frac{t}{t+a}\right)^b \tag{8.10}$$

where $c_0, c_1, \ldots c_N$, a and b are unknown regression coefficients, x_{kij} is the value of the kth parameter for the ith specimen of the jth laboratory, δ_j a random factor value that represents the deviation of jth laboratory results from the general mean, and ε_{ij} is the value for the ith specimen of the jth laboratory of a normal error term. These N parameters considered are f_R, f_L, I_c, temperature control, humidity control, support type and test configuration. The δ_j values are realizations of a random factor δ, assumed to have zero mean and normal distribution. The variance of the random factor represents variability in the data between laboratories, while the variance of the standard error term ε represents variability within laboratories.

The regression model considers quantitative variables (f_R, f_L, I_c) with their actual values, while qualitative factors, i.e. temperature control, support types, etc., are converted into binary variables (either 0 or 1). In particular, $x_1 = f_R$; $x_2 = f_L$; $x_3 = I_c$; $x_4 = 1$ if temperature was controlled and 0 otherwise; $x_5 = 1$ if the supports do not allow horizontal movements along the x-axis (Fig. 8.29), 0 otherwise; $x_6 = 1$ for load configurations with three-point bending for pre-cracking and four-point bending for creep, 0 otherwise; $x_7 = 1$ for load configurations with three-point bending for both pre-cracking and creep, 0 otherwise; $x_8 = 1$ is humidity was controlled, 0 otherwise. Temperature and humidity were assumed as qualitative variables because the time history of the values of these parameters was reported by the laboratories.

Statistical analyses were carried out independently for data on SyFRC and SFRC mixes fibres because the physical phenomena influencing the behaviour of these materials are in part different. Furthermore, different definitions of initial deformation were considered, as discussed in Sect. 9.5.1. For each case considered the following procedure was adopted. At first a model with all the parameters was fitted, outliers where then identified from the analysis of standardized regression residuals and the corresponding specimens removed from the dataset (see Fig. 8.38 for an example). The model was the refitted, considering only statistically significant parameters.

Table 8.11 reports the estimates of the parameters of the regression model as well as their p-values. Low values of these parameters indicate high statistical significance. Figure 8.39 shows the random factor values estimated for each laboratory for the different regression analyses performed.

In general, the reduction of sensitivity occurs as the initial time increases and when SFRC mixes are analysed. Figure 8.40 shows, as an example, a comparison between experimental data and regression curves for the creep coefficients at the three reference times for LAB-01 and LAB-07. The parameters for which no values are reported were found not significant in the first step of the statistical analysis. It is interesting to observe that, according to this analysis the creep coefficient for SyFRCs is correlated to f_R, I_c, temperature control, type of supports and load configuration. In particular for SyFRC, regression results suggest that the creep coefficient tends to decrease as f_R increases and to increase with I_c. Furthermore, controlling temperature has a reduction effect. The presence of supports not allowing horizontal movements is

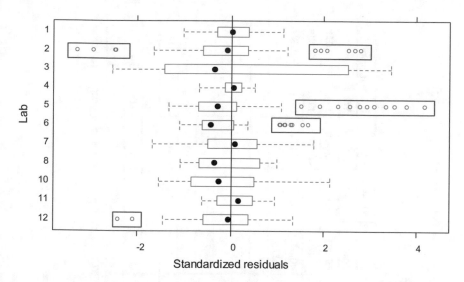

Fig. 8.38 Standardized residual obtained from nonlinear regression using the model in Eq. (8.10) and the complete dataset for MS-FRC and $t_0 = t_E$. Outliers are highlighted by red boxes

correlated to higher values of the creep coefficient. Finally, the loading configuration seems to have an effect, with 3PBT leading to higher creep values.

The statistical analysis on the data for SFRC indicates only the factors concerning the type of supports that are statistically significant. It is important to remark that the RRT was not specifically designed to fit statistical predictive equations for the creep coefficient. Therefore, the models discussed herein should not be used to this purpose and the conclusions on the significance of parameters are specific to this dataset and should not be assumed as valid in general. Furthermore, the nature of the dataset, in which many parameters change at the same time in an uncontrolled way, does not allow the exclusion of spurious correlations.

Finally, the total variability is decomposed between- and within-laboratory components. For both SyFRC and SFRC the between-laboratories variance component is higher that the within-laboratory component, i.e. the difference between mean creep coefficients of different laboratories are higher than the differences observed among the curves by each laboratory. This is typically and indicator of significant variations in the testing procedures adopted by the different laboratories.

8.5 Analysis of the Crack Opening Rate (COR) Parameter

Starting from the delayed CMOD results presented in Chap. 7 and available in the RRT database [18], crack opening rates (COR) where calculated by means of the following equation:

Table 8.11 Estimates of the parameters of the model in Eq. (8.10) and their p-values

	MS $t_0 = t_E$		MS $t_0 = t_E + 10'$		MS $t_0 = t_E + 30'$		S $t_0 = t_E$		S $t_0 = t_E + 10'$		S $t_0 = t_E + 30'$	
	Value	p-val.	Value	p-val.	Value	p-val.	Value	p-val.	Value	p-val.	Value	p-val.
c_0 (–)	1.97	<0.001	0.77	<0.001	3.77	<0.001	0.71	<0.001	1.01	<0.001	1.00	<0.001
a (days)	184.68	<0.001	252.21	<0.001	358.34	<0.001	200.20	<0.001	145.60	<0.001	136.30	<0.001
b (–)	0.26	<0.001	0.26	<0.001	0.26	<0.001	0.36	<0.001	0.41	<0.001	0.45	<0.001
c_1 (f_R) (1/MPa)	−0.76	<0.001	−0.28	<0.001	−0.76	<0.001	0.07	0.002	–	–	–	–
c_2 (f_L) (1/MPa)	–	–	–	–	–	–	–	–	–	–	–	–
c_3 (lc) (–)	0.07	<0.001	0.07	<0.001	0.04	0.001	–	–	–	–	–	–
c_4 (TC) (–)	−0.57	<0.001	−0.87	<0.001	−0.72	0.0004	–	–	–	–	–	–
c_5 (ST) (–)	3.00	<0.001	1.44	0.00	1.53	0.0008	–	–	–	–	–	–
c_6 (LC TF) (–)	−2.65	<0.001	−1.47	<0.001	−1.89	<0.001	–	–	−0.29	0.02	−0.32	0.011
c_7 (TC TT) (–)	3.80	<0.001	2.60	<0.001	1.90	<0.001	–	–	−0.13	0.35	−0.26	0.034
c_8 (HC) (–)	–	–	–	–	–	–	–	–	–	–	–	–
σ^2 between (–)	0.211	–	0.152	–	0.185	–	0.045	–	0.023	–	0.018	–
σ^2 within (–)	0.140	–	0.144	–	0.101	–	0.023	–	0.025	–	0.023	–

TC temperature controlled; *ST* support type; *LS TF* three-/four-point bending load configuration; *LS TT* three-/three-point bending configuration; *HC* humidity controlled

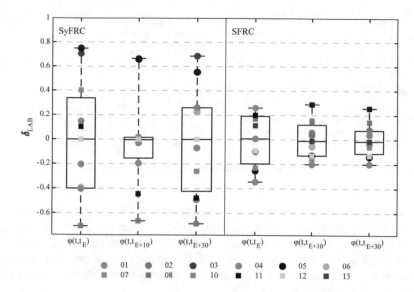

Fig. 8.39 Estimates of the random factor values (δ_j in Eq. 8.10) for each laboratory for the different models fitted

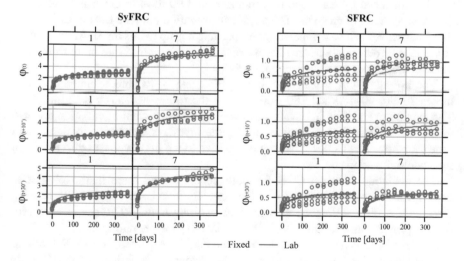

Fig. 8.40 Comparison of the predicted creep coefficient at different time lapses for LAB-01 and LAB-07 and for both types of fibres using the Eq. 8.10.

$$\mathrm{COR}^{j-k} = \left(\mathrm{CMOD}_{cd}^{k} - \mathrm{CMOD}_{cd}^{j}\right)/((k-j)/365) \qquad (8.11)$$

For this analysis, the COR results are presented in the time lapses 30–90, 90–180 and 180–360 days as reference to minimize the influence of unusual COR values.

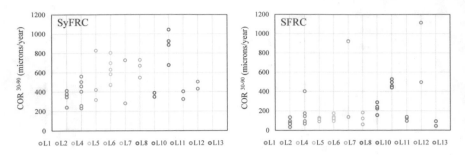

Fig. 8.41 COR$^{30-90}$ obtained from all specimens tested in flexure for both fibre types

The COR parameter is very sensitive to sudden deformations during the creep tests and sudden deformations may lead to negative displacement rates. Some laboratories reported missing values in the deformation, which implies missing COR values and discontinuity in the curves. The use of CMOD trend lines for the calculation of COR can be an alternative procedure to reduce the sensitivity to sudden deformations and the influence of missing value.

Figure 8.41 shows the COR$^{30-90}$ values for each specimen tested for both SyFRC and SFRC. The COR$^{30-90}$ parameter was adopted as a reference given that it is not highly influenced by early age deformations and 90 days in creep is a relatively common duration of creep test (90 days is the duration of creep tests in LAB-17 so that COR values can be compared for all participants up to this time). After 90 days, some delayed deformations are missing, and the shrinkage becomes more significant than the crack opening or deflection.

The average COR values for both SyFRC and SFRC are resumed for time lapses 30–90, 90–180 and 180–360 days in the following Tables 8.12 and 8.13. The maximum and minimum values, as well as the coefficient of variation are also noted. The COR decreases rapidly during the first days and trends to stabilize in time near zero. The COR$^{30-90}$ is more sensitive to the fibre type and differences in methodologies adopted.

The evolution of mean COR for both concretes and three methodologies is depicted and compared in Fig. 8.42 where outliers from LAB-06 and LAB-12 were discarded. A clear difference in behaviour between both FRC types is observed. SyFRC provides an average COR$^{30-90}$ around 450 μm/year during the first 90 days, while the SFRC gives average of 150 μm/year. As expected, the COR tends to decrease in time for most laboratories. This behaviour confirms that the SyFRC needs more time than SFRC to show stable displacement rates during the first 90 days. In this initial stage, deformation rates are significantly higher, leading to bigger differences between both SyFRC and SFRC.

A slight reduction of the coefficient of variations (CV) is achieved if the experimental COR values are normalised by the applied creep index according to the equation:

$$COR_{corr}^{j-k} = COR^{j-k} \cdot (0.5/I_c) \tag{8.12}$$

Table 8.12 Mean, minimum and maximum COR expressed in μm/year obtained from each participant from SyFRC

LAB	COR$^{30-90}$					COR$^{90-180}$					COR$^{180-360}$				
	Mean	Max	Min	SD	CV	Mean	Max	Min	SD	CV	Mean	Max	Min	SD	CV
1	354	411	239	60	17	217	262	164	38	18	106	119	90	11	10
2	401	560	230	133	33	–	–	–	–	–	–	–	–	–	–
4	521	827	316	270	52	204	292	142	78	38	104	124	90	18	17
5	609	803	468	130	21	261	379	193	66	25	205	230	172	24	12
6	504	726	281	315	63	565	808	322	344	61	111	166	55	79	71
7	649	730	548	93	14	297	324	243	47	16	264	446	162	158	60
8	368	389	348	28	8	–	–	–	–	–	85	98	72	18	22
10	881	1042	676	152	17	322	356	243	53	17	148	168	118	22	15
11	363	403	324	56	15	136	142	131	8	6	64	70	57	9	15
12	467	504	430	52	11	585	623	548	53	9	20	42	–2	31	156
13	–	–	–	–	–	–	–	–	–	–	–	–	–	–	–
11	201	224	186	21	10	103	118	94	13	13	50	58	45	8	15
16	239	243	234	6	3	122	134	110	17	14	41	47	34	9	21
12	1613	1996	1229	542	34	712	897	528	261	37	366	470	263	147	40
15	510	1271	189	415	81	273	775	126	250	92	45	148	–6	55	121
18	852	1046	706	175	21	487	811	312	281	58	199	274	154	65	33
17	3532	5844	1353	1856	53	–	–	–	–	–	–	–	–	–	–

Table 8.13 Mean, minimum and maximum COR expressed in μm/year obtained from each participant from SFRC

LAB	COR$^{30-90}$					COR$^{90-180}$					COR$^{180-360}$				
	Mean	Max	Min	SD	CV	Mean	Max	Min	SD	CV	Mean	Max	Min	SD	CV
1	81	135	34	37	46	107	165	51	48	45	29	36	18	9	30
2	159	403	68	126	79	–	–	–	–	–	–	–	–	–	–
4	112	128	91	19	17	51	57	45	6	12	27	28	27	1	3
5	130	175	94	28	21	50	105	22	32	63	40	53	33	7	18
6	713	925	137	385	54	326	466	178	145	44	41	83	19	36	88
7	122	183	61	61	50	68	122	0	62	92	–	–	–	–	–
8	227	288	156	55	24	32	52	–	28	86	–	–	–	–	–
10	480	528	440	40	8	44	80	–	55	126	–	–	–	–	–
11	120	137	95	22	18	60	68	44	14	23	36	46	27	14	38
12	805	1114	496	437	54	124	167	82	60	48	64	73	54	14	22
13	68	92	44	34	50	50	55	45	7	13	37	46	27	13	37
11	19	20	17	2	9	11	11	10	1	8	5	5	4	0	10
16	33	33	33	0	0	14	18	10	6	40	11	11	10	1	7
12	310	533	87	315	102	181	252	111	100	55	84	103	65	27	32
15	360	493	249	93	26	176	223	110	43	24	32	55	6	18	57
18	231	365	67	151	65	164	187	118	40	24	86	103	65	19	23

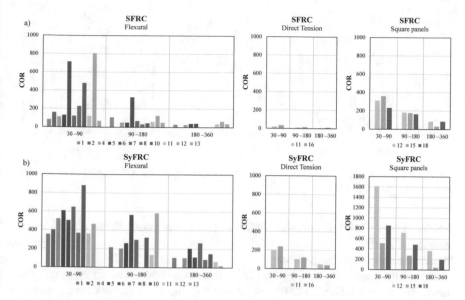

Fig. 8.42 Mean COR^{30-90}, COR^{90-180}, and $COR^{180-360}$ obtained in the different creep test for SFRC (**a**) and SyFRC (**b**)

Since most specimens are tested within an acceptable creep index range, only in those cases where the applied creep index is significantly higher than targeted by the RRT (50% $f_{R,p}$), the COR value drops according to the ratio expressed in Eq. (8.12). The higher scatter in the case of SFRC may be related to the smaller number of fibres crossing the cracked section. This is consistent with the scatter found in residual strengths $f_{R,1}$ reported by the laboratories. The CV of COR remains between 20 and 40% for the different time lapses considered, which is similar to the typical scatter found in quasi-static flexural tests of FRC. The COR is highly sensitive to small variations of the procedure and equipment adopted. Atypically high COR values are usually related to cases of broken specimens or outliers, as identified in some specimen from LAB-06 and LAB-12. Even in those cases in which high COR values were reported, the ratio between SyFC and SFRC remains constant. This demonstrates that the creep test is sensitive to the fibre type. In general, COR are consistent across most laboratories.

If the time lapse for the COR evolution calculation is kept constant (i.e. 30 days), the results is the velocity of stabilisation along the creep test duration, where a power trend line can be obtained from median values as illustrated in Fig. 8.43. This COR evolution decreases in time as the increment of delayed deformation decreases. By studying this trend line, the stabilisation of the delayed deformation could be interpreted and predicted in time. Figure 8.43 compares the COR evolution for three laboratories from different clusters and it can be observed how can differ the COR evolution between clusters. Whereas the LAB-01 from cluster A provides clear trend lines for both SyFRC and SFRC mixes, the laboratories from clusters B and D depict

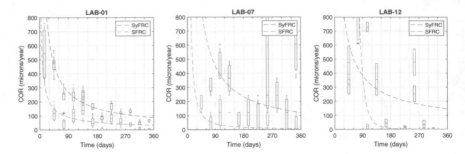

Fig. 8.43 COR evolution for different clusters: LAB-01 from cluster A, LAB-071 from cluster B and LAB-12 from cluster D

non conclusive COR evolutions curves. In addition, comparing the results within the same cluster, more homogeneous COR results and smoother curves were observed for laboratories comprising cluster A.

Part of the variability among the participants is related to the different environmental conditions during the creep test, which may induce different shrinkage deformations. Note that the shrinkage deformations become more significant at times when the delayed deformations trend to present lower values. Since one of the objectives of this RRT was to test the influence of the different creep testing procedures, it was agreed that each participant could perform the creep test in the environmental conditions available in their facilities. Notwithstanding, in the case of a future standardised test proposal, the environmental conditions during creep test should be well defined.

8.6   Influence of Fibres in the Cross Section: Fibre Counting

After the conclusion of the flexure creep test, participants were requested to count the fibres in the cracked cross-section to evaluate the fibre distribution and its influence on the creep results. Participants were instructed to divide the cracked section in three areas (top, middle and bottom) as seen in Fig. 8.44 and to count the fibres located in each of them. This procedure was agreed within the RILEM TC 261-CCF.

The information was reported in Group O of the database [18]. The fibres of both sides of the cracked sections were added for this analysis. Eight laboratories could report this additional information as seen in Fig. 8.45 where the fibre density in all the cross-section of the specimens of each participant laboratory is depicted by means of a box and whiskers plot. The average number of fibres for both mixes considering all the laboratories is also illustrated. Considering both SyFRC and SFRC mixes average number of fibres, SyFRC present 3.7 times more fibres than SFRC mixes. The bottom area of the cross-section is the most significant one since the residual performance depends directly on the fibres crossing this area. Considering only the bottom area, values between 15 and 150 fibres were found for some specimens, nearly 10 times

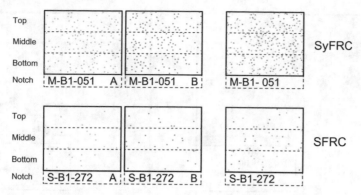

Fig. 8.44 Fibre counting in different areas of cross section in one sample for both SyFRC and SFRC

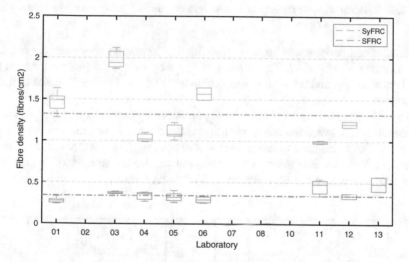

Fig. 8.45 Fibre density in the whole cross-section obtained from the flexure creep test

more fibres depending on the mix. Steel fibres showed a fibre density in the bottom of the cross-section between 0.23 and 0.6 fibres/cm^2 whereas macro-synthetic fibres showed a density between 0.8 and 2.26 fibres/cm^2 at the same area. The scatter in the results is particularly high for SyFRC since fibre counting is rendered more difficult by the fact that synthetic fibres broken during the flexural test may be counted twice.

The influence of the number of fibres in the residual strengths f_L and $f_{R,1}$ has been analysed in Fig. 8.46. Since the residual strength at LOP (f_L) depends on the concrete matrix strength, a poor correlation was observed. On the contrary, a better correlation was expected to residual strength $f_{R,1}$ since the number of fibres in the cross section directly influence the residual behaviour of FRC.

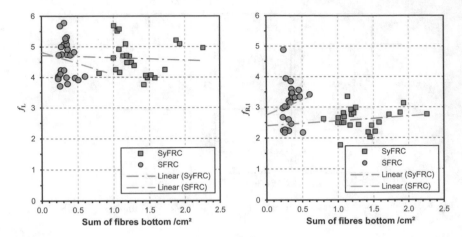

Fig. 8.46 Residual strengths f_L and $f_{R,1}$ versus number of fibres in the bottom cracked cross section

Similar comparison was made with the total deformation after one year of creep test as shown in Fig. 8.47. The number of fibres by square centimetre for both the total cross section and the bottom were compared and good correlation was found in both cases.

Regarding the influence of fibre density on the creep coefficients, Fig. 8.48 shows the correlation between the number of fibres at the bottom of the cross-section and the creep coefficient referred to creep stage at 360 days ($\varphi_{c,w}^{360}$). SFRC and SyFRC show considerable differences in terms of the correlation of both parameters. In general, a poor correlation between fibre density and $\varphi_{c,w}^{360}$ was found.

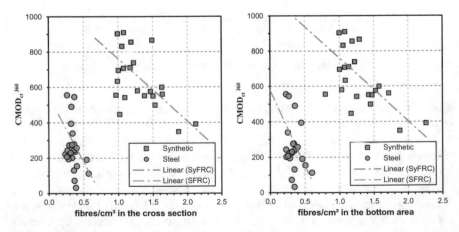

Fig. 8.47 Correlation between number of fibres in the cross section and total delayed deformation at 360 days

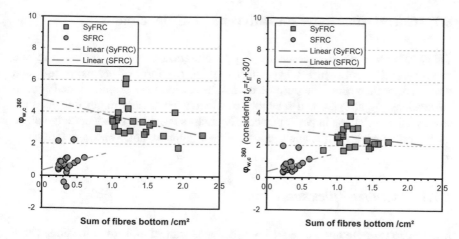

Fig. 8.48 Creep coefficient in creep stage versus number of fibres in cross section

The correlation between COR [μm/year] and fibre density at the bottom of the cross-section was investigated for the interval 30 to 90 days (COR^{30-90}), 90 to 180 days (COR^{90-180}) and 180 to 360 days ($COR^{180-360}$), as depicted Fig. 8.49. Again, SFRC and SyFRC mixes show considerably different results. No clear correlation was found. The reason for the poor correlation between fibre content in flexural zone of the beams and creep parameters could be the absolute low level of creep starting at 0.5 mm pre-crack opening and a load level of 50% of the short-term load. Moreover, the scatter induced by the differences in the testing procedures could hinder the identification of the correlation.

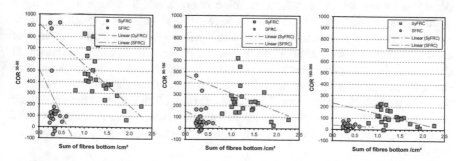

Fig. 8.49 COR^{30-90}, COR^{90-180} and $COR^{180-360}$ [μm/year] versus number of fibres in the bottom of the cracked cross section

8.7   Correlation Between Different FRC Specimens

This section will discuss the nature of possible correlations in post-crack tensile creep behaviour between FRC specimens taking the form of beams, panels, and direct tensile members. The purpose of this discussion is to examine whether such correlations might exist, and secondly consider what factors might influence the associated correlations in performance. The factors requiring consideration include member thickness, friction and supports, fibre alignment, notch and variability.

8.7.1   Member Thickness

Most creep tests are undertaken using specimens subject to flexural loading. These usually take the form of simply supported beams, or panels. Examples include beams based on EN 14651 [22], ASTM C1609/C1609M [50], or panels based on EN 14488.5 [28] or ASTM C1550 [29]. The main alternative to flexural specimens is direct tension specimens such as FRC notched cylinders or dog-bone coupons with a square cross-section (CNR DT 204). In addition, single fibre direct tension tests can be undertaken with the fibre either embedded in paste/concrete or not. The question of how results obtained from these different types of tests are related is the subject of the current discussion.

Fibres used to reinforce concrete are of finite length and typically suffer a pull-out based mode of post-crack behaviour. This means that the length of fibre resisting pull-out falls as a crack widens, and thus resistance to pull-out usually falls (on average) as a crack widens. Simple flexural members such as standard beam and panel test specimens will exhibit a variation in crack width through the thickness of the member, and thus the resistance to fibre pull-out will vary through the cross-section. This means that members of differing thickness will exhibit different flexural resistance characteristics, even when the same FRC mixture is used to produce all the specimens. If a simple elastic stress profile is used to model post-crack residual strength, the apparent flexural strength will vary for a given FRC mixture even for the same maximum crack width. This makes it difficult to compare the post-crack performance of flexural specimens with different thicknesses. All the most common standard flexural test specimens have a different thickness (ranging from 75, 100, 125 to 150 mm).

To avoid these problems the flexural stress profile through a specimen can be estimated and used to back-calculate a direct tensile stress profile. This will make it possible to compare flexural specimens of differing thickness and compare flexural specimen performance to that of direct tension specimens. This may sound simple, but back-calculating the equivalent direct tension stress profile through a FRC section is subject to many assumptions and introduces uncertainties that are related to the assumptions made about the shape of the profile. Despite this, back-calculating the direct tensile strength and expressing this as a function of crack width is the only way flexural performance from specimens of differing thickness and direct tension test specimen data can truly be compared.

The discussion outlined above applies to short-term quasi-static testing of FRC. Long-term investigations of creep may or may not be similarly affected by specimen thickness and the shape of the tensile stress profile through the specimen section. This is because creep performance is normally expressed in terms of long-term time-dependent crack width increase relative to short-term crack width measured for the same specimen. It is likely that the relationship between applied load, stress distribution through the depth of a section, and the relative extension or relaxation of fibres as stress is re-distributed through a section with the passage of time, are all very complex. Factors affecting behaviour are likely to include: how a sustained tensile load affects the rate of fibre pull-out or elongation at a crack, the location of each fibre through the depth of a section, the orientation of each fibre relative to the direction of tension, and the re-distribution of load between fibres over time as some pull-out more than others. Given the confluence of all these factors, it is very difficult to predict the relative rates of creep in different types of test specimen even when the same mixture is used to produce all of them. A simpler approach is simply to measure creep behaviour in numerous different types of specimen made using the same mixtures.

8.7.2 Friction and Supports

Beam and panel specimens both rest on supports during a test, and the friction experienced at the support will influence the apparent residual flexural strength of the FRC. Since it is effectively impossible to completely eliminate friction at the supports, the true flexural and direct tensile performance of an FRC mixture will always be less than the apparent performance. The proportion of the apparent residual strength that is attributable to friction at the supports will vary with the geometry of the test configuration, the possible presence of rollers and their effective Coefficient of Friction, the thickness of the specimen, and other factors. If a correlation is sought between the post-cracking performances of a particular FRC mixture assessed using different types of specimen, then the effect of friction must be isolated and corrected for so the true performance of the FRC can be distinguished from the apparent performance.

In the context of quasi-static short-term testing of beams, some authors examined how friction in supporting rollers can exaggerate the flexural post-crack performance of FRC beams [51, 52]. In the ASTM C1609/C1609M beam test, the apparent post-crack flexural residual strength of a mixture can be up to 3 times greater than the true flexural strength if locked-up supporting rollers are used. This is because the Coefficient of Friction (CoF) between steel and concrete is about 0.75–0.80, while the CoF for a properly functioning roller is about 0.10. Since many roller designs appear susceptible to 'lock-up' when a vertical load is applied, standard practice ASTM C1812 for supporting roller design [53] was recently introduced and is now prescribed in ASTM C1609/C1609M. This type of roller has a reproducible CoF equal to 0.10 and infinite travel. When the CoF is 0.10, about 10% of the apparent residual flexural strength of a FRC beam is attributable to friction. Unfortunately, EN 14651 has no similar requirements placed on supporting rollers and thus the proportion of apparent performance measured in this test that is attributable to friction is unknown (and most likely differs between laboratories).

The ASTM C1550 [29] panel test includes flat pivoted steel plates under each of the three support points. The CoF for this design of supporting plate has been measured [54] to be 0.75–0.80, with the result that about 12–13% of the apparent performance of each panel tested using this method (both energy absorption and residual flexural strength) is attributable to friction.

The EN 14488-5 [28] square panel test is subject to a great deal more friction than any of the other commonly used post-crack performance tests. Flexure test conducted on square panels [55] and Norwegian round panels [56] placed on standard continuous supports were compared to similar tests conducted on Teflon supports and a difference in energy absorption of up to 35% was found. Unfortunately, specifications for the design for the supports in EN 14488-5 [28] are vague, and inspection of test machines in numerous laboratories has shown wide variations in interpretation of the specification. Most laboratories use a solid steel square base without any rollers. When a cracked square panel slides over this type of support, very high levels of friction arise at the boundaries giving rise to high energy absorption associated with friction. Since the crack pattern typically varies between panels depending on how flat the base is, the degree of friction experienced among a group of nominally identical specimens is unlikely to be uniform. Given both the very high proportion of apparent post-crack performance attributable to friction in this test, and the variability in this proportion, correlations in post-crack performance between this test and other test methods are likely to be problematic.

The effects of friction on short-term quasi-static testing of FRC are only slowly being explored by the FRC community. The possible influence of friction on creep assessment has not been examined to date. However, one can estimate that non-rolling support rollers will have a significant effect on the apparent creep behaviour of cracked FRC beams. This effect is difficult to estimate if the time dependent CoF exhibited by a roller is not known. It is recommended that an investigation of long-term slippage across a fixed steel roller or flat plate be undertaken to identify the magnitude of the time-dependent CoF at the concrete-steel boundary.

8.7.3 Fibre Alignment

The apparent post-crack performance of a FRC mixture is dependent on the mean orientation of the fibres contained in the mixture. It is commonly assumed that the fibres are uniformly and randomly distributed and oriented, but this is seldom true. Many factors, including the way in which the specimen was cast, can give rise to a biased distribution of fibres both in location and orientation. Most fibres generate the highest resistance to pull-out when orientated perpendicularly to a crack, so if some process were to lead to a bias in the orientation of fibres within a specimen so that the majority became aligned with the direction of principal tensile stress, then the apparent performance of the mixture will be higher than if the fibres were truly randomly oriented. Such a bias can most readily occur when long fibres are cast in a narrow mould or in flowing concrete [57].

Alignment of fibres can also occur due to the so-called 'boundary effect'. The hard surface of a mould will not let a fibre intersect this boundary, unlike a free cast surface. Fibres near a hard mould surface will therefore become aligned with this surface. The effect is made worse by vibration as this induces fibre alignment more deeply within the concrete than for non-vibrated concrete. Beam moulds are narrow and channel concrete during casting compared to panel moulds. Beams made with flowable concrete containing long fibres that are subsequently vibrated are the most severely affected by fibre alignment. A FRC beam specimen made this way is highly likely to exhibited higher post-crack flexural performance than panels made with the same concrete mixture. It would therefore be difficult to establish a true and representative correlation in post-crack performance between beams, panels, and direct tension specimens made in this way.

One possible solution to this problem is to cut all specimens from a large slab of hardened FRC and to pay careful attention to the possible alignment of the fibres with the lower boundary. The fibres within the various specimens are then likely to display a similar degree of anisotropy. The question of whether fibre alignment within test specimens affects apparent creep behaviour remains unresolved, but if correlations in creep behaviour between different types of specimens are to be examined then this issue requires consideration.

8.7.4 Notch

The presence of a notch on the tensile face of a test specimen will normally lead to a crack forming at the location of the notch in preference to any other location. When no notch is present, the crack tends to form at the weakest location within the local region of uniform tensile stress. If the mean performance of a large group of notched specimens is compared to the mean performance of a large group of un-notched specimens of identical composition, it is generally noticed that the notched specimens exhibit higher tensile performance. In addition, as the crack propagates through the section, it tends to divert around the ends of fibres when the tensile stress field is uniform but tends to intersect fibres when the tensile stress field is narrow [58]. This means that notched specimens tend to exhibit both a higher tensile strength at first crack and in the post-crack range than specimens lacking a notch. This effect will clearly influence the nature of any correlation in performance between notched and un-notched specimens.

Since creep is normally assessed in terms of the performance of a specimen in the long-term relative to its short-term quasi-static performance, the presence or otherwise of a notch will influence both the short and long-term performance. It is therefore not clear whether a notch will influence possible correlations in the creep behaviour of notched compared to un-notched specimens.

8.7.5 Variability

If a correlation in performance between panels, beams, and direct tension specimens is to be identified, the confidence possible in the correlation will depend on the degree of variation evident in the mean performance obtained for each specimen type. In the context of short-term quasi-static testing, the within-batch Coefficient of Variation in post-crack performance is related to the number of fibres crossing the crack in the FRC specimen, the degree of randomness in location and orientation, variations in unconfined compressive strength (UCS) within the specimen, and factors such as friction at the supports and the geometry of the specimen. The summed outcome of all these factors is that large specimens such as panels tend to produce post-crack performance results that exhibit a lower CV than smaller specimens such as beams and direct tension coupons. The confidence interval on the correlation parameters relating the performance of different types of specimen is most affected by the specimen set exhibiting the largest magnitude of CV in performance.

Given that creep is commonly assessed as a ratio of long-term performance (often expressed in terms of crack width) relative to short-term performance measured for the same specimen, it is not immediately apparent that variability in creep performance is affected by the size of the specimen in the same way that variability in quasi-static performance is affected.

8.8 Influence of Ageing of Concrete Matrix on FRC Characterization Test and Creep Tests

This section discusses the possible effect of ageing on the flexural and tensile creep behaviour of FRC specimens. With time, cement hydration changes the characteristics of the fibre-matrix interfacial transition zone. Ageing can also be associated with weathering if exposure conditions are suitable.

8.8.1 Embrittlement

Embrittlement is a phenomenon by which the ductile high-energy pull-out mode of fibre post-crack behaviour characteristic of young concrete changes to a brittle low-energy rupture-based mode of failure as a result of a steady increase in bond strength between fibre and cementitious paste over time. The fibre anchorage essentially becomes so strong that the pull-out resistance exceeds the tensile strength of the fibre. This phenomenon has been found in SFRC and some SyFRC mixes when overly strong concrete mixes are used together with fibres of insufficient tensile strength. The transition from a non-embrittled to embrittled state may occur many months after casting, depending on the strength and cementitious characteristics of the concrete matrix.

Evidence from different studies [59, 60] indicate that certain hooked-end steel fibres are sensitive to this process, due to their relatively limited tensile strength of 1100 MPa. However, the minimum UCS (at the time of testing) required to initiate a transition from pull-out to rupture of certain steel fibres is 45 MPa. A recent investigation [61] indicates that SFRC incorporating certain fibres with a UCS less 45 MPa will not transition to an embrittled state within 10 years after casting. The SFRC mixes used in the current RRT did not exceed 38 MPa by 500 days. In addition, the post-crack residual strength at CMOD 2.5 mm ($f_{R,3}$) depicted in Fig. 3.3 indicates no systematic fall in performance at 500 days. This appears to confirm that no embrittlement has occurred in the SFRC specimens tested in the RRT.

Most macro-synthetic fibres do not appear susceptible to embrittlement [62], but this has not been confirmed for the macro-synthetic fibres used in the present study. Nevertheless, the record of post-crack flexural strength $f_{R,3}$ indicates no loss of performance at 500 days, suggesting that embrittlement is unlikely for the SyFRC mixtures tested in the RRT.

The embrittlement is of significance to creep testing is that an embrittled mixture is unlikely to exhibit time-dependent pull-out of fibres, and thus crack widening would primarily become a function of fibre elongation with time. In contrast, the fibres in a mixture that has not transitioned to an embrittled state would exhibit both elongation and progressive slippage relative to their initial anchorage depth. Crack widening due to tensile creep is therefore likely to be slower in an embrittled FRC mixture than in a non-embrittled mixture.

8.8.2 Corrosion

The primary manifestation of tensile creep in cracked FRC members is a steady widening of cracks with the passage of time. If the cracks become sufficiently wide, and the fibres are sensitive to corrosion, then corrosion might occur should conditions be suitable. This is fully possible in external structures subject to weathering but is highly unlikely in most indoor laboratory environments. Given the benign conditions prevalent in most laboratory environments, it is considered unlikely that creep and corrosion will occur concurrently to the extent that specimen performance is affected by corrosion to any significant extent.

Chapter 9
Statements and Conclusions

Aitor Llano-Torre, Pedro Serna, Sergio H. P. Cavalaro, Nicola Buratti, E. Stefan Bernard, William P. Boshoff, Raúl L. Zerbino, Hans Pauwels, Wolfgang Kusterle, Emilio Garcia-Taengua, Rutger Vrijdaghs, Clementina del Prete, Karyne F. dos Santos, Benoît Parmentier, and Claudio Mazzotti

Abstract A round-robin test (RRT) on creep behaviour in cracked section of fibre-reinforced concrete (FRC) was realised with the participation of 19 international laboratories from 22 institutions. The same concrete matrix was used to design two different FRC mixes with steel (SFRC) and macro-synthetic (SyFRC) fibres and to

A. Llano-Torre (✉) · P. Serna
Institute of Concrete Science and Technology ICITECH, Universitat Politècnica de València (UPV), Valencia, Spain
e-mail: aillator@upv.es

P. Serna
e-mail: pserna@cst.upv.es

S. H. P. Cavalaro
School of Architecture, Building and Civil Engineering, Loughborough University, Loughborough, UK
e-mail: s.cavalaro@lboro.ac.uk

N. Buratti · C. del Prete · C. Mazzotti
Department of Civil, Chemical, Environmental and Materials Engineering DICAM, University of Bologna, Bologna, Italy
e-mail: nicola.buratti@unibo.it

C. del Prete
e-mail: clementina.delprete2@unibo.it

C. Mazzotti
e-mail: claudio.mazzotti@unibo.it

E. S. Bernard
TSE Technologies in Structural Engineering Pty Ltd., Sydney, Australia
e-mail: s.bernard@tse.net.au

W. P. Boshoff
Faculty of Engineering, Built Environment and Information Technology, University of Pretoria, Pretoria, South Africa
e-mail: billy.boshoff@up.ac.za

R. L. Zerbino
LEMIT-CIC and Faculty of Engineering UNLP, La Plata, Argentina
e-mail: zerbino@ing.unlp.edu.ar

© RILEM 2021
A. Llano-Torre and P. Serna (eds.), *Round-Robin Test on Creep Behaviour in Cracked Sections of FRC: Experimental Program, Results and Database Analysis*, RILEM State-of-the-Art Reports 34, https://doi.org/10.1007/978-3-030-72736-9_9

produce 451 FRC specimens. The use of 124 specimens for creep tests for one year following four main methodologies provided a huge database with more than 15,000 data. The main statements and conclusions derived from the RRT database analysis are summarised in this section. Note that the conclusions derived from this RRT are limited to the ranges of parameters, procedures and variables considered here and should not be extrapolated to other cases.

A comprehensive round-robin test (RRT) on creep behaviour in cracked section of fibre-reinforced concrete (FRC) was realised with the participation of 19 laboratories from 22 institutions including the most experienced ones in the world on this creep procedure in both at research or control level, representing the technical and scientific state of the art on this topic. This report presents the organization, procedure, and result analysis of the international RRT focused on long-term behaviour of FRC specimens in their cracked state. The experimental program, concrete mixes and materials, different creep testing methodologies, criteria applied for the test parameters definition, general procedures and the analysis of the creep test results have been described and justified.

Assuming the big variability on the proposed test methodologies reported in the bibliography, the first part of the work consisted of grouping the different methodologies in four main groups and analysing its actual relative significance. Therefore, four testing procedures were selected as the most representative of the actual state of the research and applications: flexure in prismatic specimens, direct tension, flexure in square panels and flexure in round panels. The flexure creep procedure in prismatic specimens was developed inspired in EN 14651 or other flexural standards for the

H. Pauwels
NV BEKAERT SA, Zwevegem, Belgium
e-mail: hans.pauwels@bekaert.com

W. Kusterle
OTH Regensburg, Regensburg University of Applied Sciences, Regensburg, Germany
e-mail: wolfgang@kusterle.net

E. Garcia-Taengua
School of Civil Engineering, University of Leeds, Leeds, UK
e-mail: E.Garcia-Taengua@leeds.ac.uk

R. Vrijdaghs
Building Materials and Building Technology Section, KU Leuven, Leuven, Belgium
e-mail: rutger.vrijdaghs@kuleuven.be

K. F. dos Santos
Department of Civil and Environmental Engineering, University of Brasília, Brasília, Brazil
e-mail: karyne.ferreira@aluno.unb.br

B. Parmentier
BBRI Belgium Building Research Institute, Limelette, Belgium
e-mail: benoit.parmentier@bbri.be

analysis and characterization of the FRC material as general concept, whereas the square panel and round panel flexure creep test procedures were developed thinking in tunnel applications. Additional tests were also performed during the RRT like shrinkage and creep in compression test.

The main objectives of this RRT were to detect differences between methodologies and to verify the repeatability and reproducibility of different test methods. The consecution of these goals would provide the basis for a proposal of standard test conditions based on the RRT conclusions and create a consensus on how creep on cracked FRC should be evaluated for structural applications. Therefore, the same concrete matrix was used for two different FRC mixes: one with steel fibres (SFRC) and another with macro-synthetic fibres (SyFRC). Only the superplasticizer was adjusted to guarantee the same workability for all FRC batches. The fibre dosage was chosen to ensure a similar residual strength $f_{R,3}$ in both FRC mixes regardless of the fibres. Note that this limited choice of concrete matrix and fibre cannot be considered representative of the wide variety of FRC applied in practice.

Different responses on creep were obtained depending on the fibre material, with higher delayed deformations and creep coefficients for SyFRCs. Nevertheless, results should not be extrapolated since the market of fibres is extremely diverse with a big number of brands, materials qualities and fibres shapes that can lead to very different behaviours.

General testing procedure and parameters were discussed and defined by the RILEM Technical Committee 261-CCF before the beginning of the RRT to coordinate similar testing criteria. This agreement included the definition of main concepts and a series of test parameters like the reference stress considered as the residual capability of the concrete ($f_{R,p}$), the creep index applied during the creep period (I_c), and the initial pre-crack opening before the creep test ($CMOD_p$). In order to reach measurable creep deformations, the agreed creep index or stress level was $50\% f_{R,p}$ at crack mouth opening displacement (CMOD) of 0.5 mm for flexural and crack opening displacement (COD) of 0.2 mm for direct tension tests, whereas for square and round panel tests was $60\% f_L$. Extrapolations of conclusions described here to conditions different from those considered here should be avoided.

Besides the basic FRC characterization tests, as result of the RRT, a huge database [18] with a total of 124 specimens tested in creep in cracked state has been created. The resulting database reports an average of 140 data from each specimen, leading to more than 15,000 data.

The database [18] includes for each specimen all the aspects related to testing details, the laboratory environmental conditions, testing procedure, and CMOD, COD or δ deformations evolution during all the testing stages: pre-cracking, creep period and post-creep test until failure. All these data will allow the traceability of all testing details. A detailed description of the specific procedures adopted by each laboratory as well as equipment and main variables may be found in Chap. 5.

This report presents an initial analysis of these results and leaves to other authors the opportunity to expand the analysis and compare with their own results. The final RRT database [18] was published under a Creative Commons license as supplementary material of this RRT State-of-the-Art Report and it is available for the scientific community to improve the global knowledge in the long-term deformations of the cracked FRC specimens.

9.1 Statements

This section summarises the main statements derived from the RRT, which are the basis for the conclusions. The statements are organised following a similar structure than the RRT Report document.

Regarding the analysis of the main variables and parameters of the RRT analysed in Sect. 8.1, the following statements are exposed:

- More than 14 tons of FRC and 451 specimens were produced with the same composition and components in the same location to mitigate scatter related to the production. The preliminary tests confirm similar characteristics of all mixes produced in the scope of the RRT.
- A total of 124 specimens were tested in creep for 360 days: 86 in flexure, 10 in direct tension, 24 square panels and 4 round panels. The most extended methodology is flexure in beams (69% of the specimens tested), followed by square panel test (19%), direct tension test (8%) and round panel test (3%).
- A high variation of environmental conditions was found amongst the laboratories. Two main variables were controlled: temperature and relative humidity.
- Twelve laboratories reported that creep tests were performed in temperature-controlled conditions whereas four performed creep tests registering the temperature fluctuations experienced during the test. Those laboratories that did not control the temperature present higher scatter in temperature values.
- The variability in the relative humidity among participants is significantly higher than observed for the temperature. Only ten laboratories reported a control of the humidity. However, even in those cases where controlled humidity was reported, significant humidity variations were found over time.
- The average pre-cracking level values obtained for beams tested in flexure, round panels tested in flexure and specimens tested under uniaxial tension are consistent and close to the target value defined at the beginning of the RRT. In the case of square panel, the pre-cracking criteria were different by each laboratory, which led to values ranging from 800 to 3000 microns.
- The creep index (I_c) or stress level is defined as the ratio between the applied stress during creep stage ($f_{R,c}$) and the residual strength at $CMOD_p$ ($f_{R,p}$). Depending on the methodology, different deviations were found regarding the applied creep index. In the case of flexural creep test, the higher deviations were found in those laboratories performing multi-specimen setup, whereas in the case of square panel tests, very scattered creep index values were obtained due to significant differences in the procedure. Considering all methodologies, most specimens are in the range from 45 to 65% of the $f_{R,p}$. Only in the case of LAB-15 creep index values differ

from this threshold, due to a different pre-cracking criterion, ranging from 95 to 135% of F_L.

The analysis of the flexural creep test results on prismatic specimens exposed in Sect. 8.2.1 led to the following statements:

- The flexural creep test is the most widely adopted procedure in the RRT as occurs in the scientific literature. It is a relatively easy procedure to perform and it may be integrated into the general laboratory experience. This procedure used prismatic standard specimens as in the standard characterization tests of the FRC post-cracking properties. Despite the extended use of flexural creep test procedure, the results of this RRT show a high variability in terms of delayed CMOD deformations between the laboratories that adopted this test.
- An exponential basic curve relating the evolution of delayed deformations over time was found from the statistical analysis when the influence of the methodology and the dependence on the creep index and the concrete residual strength were isolated. The proposed basic CMOD curve $F_{(t)}$ is a function of time t (days) and depends on the fibre type and other parameter affecting concrete creep.

$$F_{(t)} = a \cdot t^b$$

- The ratio between the basic curve for both SyFRCs and SFRCs ranges from 2.2 at the beginning of the creep test to 3 at 360 days.
- Other dependencies can be decoupled into univariate functions that are multipliers of the basic curve. The multiplying functions account for the influence of the stress applied in MPa $F(f_{R,c})$ and the creep index $F(I_c)$ expressed in percentage (%).

$$CMOD = K \cdot F(I_c) \cdot F(f_{R,c}) \cdot F(t)$$

- One additional multiplier appears in the previous equation as a constant K which includes the differences introduced by the different procedures and depends on the laboratory.
- The statistical analysis it shown that laboratories using similar procedures present similar results, which led to the following classification in clusters. This clustering classification may support the definition of a standard testing procedure as it revealed which procedures are similar and comparable.

  - Cluster A: LAB-01, LAB-02, LAB-04, LAB-05, LAB-06, LAB-08 and LAB-11.
  - Cluster B: LAB-07 and LAB-10.
  - Cluster C: LAB-03.
  - Cluster D: LAB-12.
  - Cluster E: LAB-13.

- The Cluster A is composed by seven laboratories with similar test methodology and setup. Most used a three-point bending test (3PBT) setup for pre-cracking tests and a four-point bending test (4PBT) in a multi-specimen setup configuration during the creep test with lever arms introducing the lead in the creep frame. The Cluster A represents 70% of the specimens tested in flexure creep methodology.

- The Cluster B consists of two laboratories that performed 3PBT in both pre-cracking and creep tests in a single specimen setup. The Cluster B represents 16% of the specimens tested in flexure creep.
- The Clusters C, D and E are composed by one laboratory each, what implies significant differences regarding the procedure adopted. Although these laboratories used a 4PBT configuration for both flexure pre-cracking and creep tests, they have differences in terms of notching, specimen size or load configuration.
- Significant differences in methodological approaches among laboratories were detected, which is reflected in the K parameter depending on the cluster. This leads to differences of up to 4 times in terms of CMOD assessed by laboratories in different clusters. Such behaviour is observed regardless of the fibre material.
- This clustering classification may be useful in order to a further definition of a standard testing procedure, since it revealed which procedures are more similar and comparable.
- The conclusions and the basic creep curve equation obtained from the analysis of the results in this RRT are limited to the ranges of variable studied. Extrapolations are not recommended.

The analysis of the direct tension test results in Sect. 8.2.2 supported the following statements:

- Direct tension test is a more complex and unstable procedure with a less extended use in research and practice. This RRT includes 10 specimens tested following the direct tension creep test procedure. The small number of specimens tested limits the conclusions derived from this study.
- As in the case of the flexural creep test of prisms, the results from the direct tension creep test fit the exponential equation:

$$COD = a \cdot t^b + c$$

- In the direct tension creep tests, the ratio between delayed COD of the SyFRC and SFRC ranges from 2.8 at the first days of test to 4.0 at 360 days. Both are bigger than the ratio observed in the flexural creep test of prisms.
- Since the number of participants is limited and the procedure control is higher, the influence of parameters like creep index, concrete strength or testing procedure was not assessed.

The analysis of the results from square and round panel creep tests presented in Sect. 8.2.3 yield the following statements:

- Square panels test based on the EFNARC or Asquapro methodology and EN 14488-5 standard were performed in the RRT with an important contribution of 24 specimens tested in creep.

- The structural redundancy of the square panel setup compromises the interpretation of results. This method can provide valuable information for sprayed concrete applications in underground construction where the method is widely used for quality control.
- Significant differences regarding to pre-crack level, creep index and applied load were found between the three participant laboratories. This compromised the comparative analysis or the evaluation of the advantages of this testing procedure.
- In the case of square panel creep tests, the ratio between the differed displacements of the SyFRC and SFRC mixes is quite scattered ranging from 1.15 to 6.8 at 360 days depending on the laboratory.
- During the creep tests, one square panel of each fibre material tested by LAB-18 failed during the creep tests between 120 and 150 days. This failure may be caused by the 3 mm pre-crack level used by the participant.
- Four round panels creep test based on ASTM C1550 standard were also performed in the RRT in four SyFRC specimens. Unfortunately, the short duration of creep tests and the high variability in the delayed deformations, compromised the comparison between methodologies.

The analysis of differences in the procedures followed by each laboratory in the flexure creep test of beams (Sect. 8.3) led to the following statements:

- Five different load configurations were detected for flexure creep test methodology in prismatic specimens, which demonstrates the significant variability in methodologies adopted.
- Despite this variability in load configuration, most participants perform the same load configuration combination consistent with Cluster A (3PBT load configuration for pre-cracking and 4PBT for creep test).
- The previous statement is linked to the use of multi-specimen configuration, which generally implies a 4PBT to ensure more stability in the column of specimen. The change from a 3PBT load configuration for pre-cracking to a 4PBT load configuration for the creep test requires the recalculation of the load to be applied by means of the stress equations.
- The multi-specimen setup is an effective way of increasing the number specimens tested and reducing time and costs. However, in some cases where the multi-specimen configuration is used, a higher scatter in creep index values was obtained due to the challenge to keep the same creep index for all the specimens in the setup.
- Participants with higher scatter in terms of creep index usually also present a higher scatter in the delayed CMOD displacement.
- In some multi-specimen creep frame configurations, specimens are kept independent from others by means of steel plates. These plates reduce the impact of the collapse of one specimen on the others from the same column during the creep test and provide continuity.
- Five different types of support or loading rollers were used by the laboratories performing flexure creep test in beams, which highlights the lack of standardisation in the methodology. Such a variation in the boundary conditions may

affect the creep test results. The support boundary conditions types were classified depending on three degrees of freedom allowed.

Considering the joint analysis of all creep test methodologies in the RRT (Sect. 8.3) and the analysed differences between procedures followed by each participant, the following statements can be exposed:

- The use of lever arm is the most extended and confident way of applying the load in the creep frames (twelve laboratories used lever arms, four laboratories used a hydraulic jack, one laboratory used screwed bars and one laboratory used dead load directly applied to the specimens).
- A similar methodology was detected, and more similar and repetitive results were obtained in laboratories where a lever arm was used. The use of an upper lever arm (i.e. direct application of load to the specimen) is the most extended approach, being used by seven laboratories. Conversely, five laboratories used a lower lever arm configuration (indirect application of load by means of tensile bars).
- Considerable differences regarding the duration of the loading stage of the creep test (t_{ci}) were detected. Laboratories were classified depending on t_{ci} in four groups as follows:

  - Low: $t_{ci} < 1$ min—seven laboratories.
  - Medium: $1 < t_{ci} < 6$ min—four laboratories.
  - High: $6 < t_{ci} < 15$ min—three laboratories.
  - Very High: $t_{ci} > 15$ min—two laboratories.

- Temperature is clearly easier to control during the creep test. Approximately 83% of the participants informed that the temperature was controlled.
- Only 75% of the participants controlled the relative humidity during the creep test, although the scatter in the relative humidity was significantly higher than that of the temperature. Despite such controlled humidity conditions, high scattered values were registered. Therefore, the relative humidity control must be more exhaustive due to the influence of humidity in shrinkage.

Two main parameters from the RRT were evaluated: the creep coefficient (φ) defined as the ratio between delayed and instantaneous deformations, and the crack opening rate (COR) defined as the average crack opening velocity (in microns/year) in certain lapses of time. The analysis of these parameters derived from flexure creep tests in Sects. 8.4 and 8.5 support the following statements:

- The creep coefficient referred to the creep period defined as the ratio of the delayed deformations at certain times ($CMOD_{cd}{}^{j}$) to the instantaneous deformations ($CMOD_{ci}$) was evaluated. The instantaneous deformation $CMOD_{ci}$ as well as the short-term deformations $CMOD_{ci}{}^{10'}$ and $CMOD_{ci}{}^{30'}$ were used for creep coefficient calculation and analysis.
- Creep coefficients referred to creep stage obtained for both mixes from specimens tested on flexure creep test were compared and analysed. SyFRC mixes provided higher and more scattered creep coefficients referred to creep period than

SFRC mixes. Moreover, SyFRC specimens are more sensitive to the instantaneous deformation definition and the time in which the load is applied parameters.

- Outliers were detected and removed from the statistical analysis to find correlations between some quantitative variables (f_R, f_L, I_c) and the qualitative factors such as temperature or humidity control, support types, load configuration.
- The creep coefficients referred to origin of deformations were obtained using as instantaneous deformation the $CMOD_{ci}$ deformation including the residual crack opening $CMOD_{pr}$ after the pre-cracking test. The creep coefficients were included in the RRT database for ulterior analysis (see Appendix D for time evolution). Whenever it was possible, creep coefficients referred to origin were depicted considering the three different definitions of instantaneous and short-term $CMOC_{ci}$ used in this RRT.
- COR parameter was evaluated in three different time lapses: COR^{30-90}, COR^{90-180} and $COR^{180-360}$. Results confirm that COR decreases with time.

The following statements are derived from the analysis of the influence of the fibre count in the cracked cross-section in Sect. 8.6:

- The SyFRC fibre count shows higher variability than the SFRC fibre count. This may be explained by the bigger difficult in the counting process of macro-synthetic fibres, some of which break during the flexure post-creep test and could be counted twice.
- The average number of fibres of SyFRC is approximately 10 times bigger than the average number of fibres of SFRC. This is attributed to the difference in density of the fibre materials. The bigger number of fibres allowed a clearer identification of correlations in the case of the SyFRC.
- A correlation was found between the SyFRC fibre count and the φ and COR for different time lapses: the higher number of fibres in the cross-section, the lower φ and COR is obtained, which indicates less deformation over time.

9.2 Conclusions

Considering the previous statements, the main conclusions arising from the RRT are:

- The testing procedure used by laboratories in Cluster A for flexure creep tests seems to be the most reliable and extended. Most laboratories used the same load configuration (3PBT for pre-cracking tests and 4PBT for creep test).
- Although the use of multiple beam test set-up is more extended due to space save in the facilities, from a technical perspective, more reliable results could be obtained in a single beam set-up. Flexure tests performed with a single specimen per creep frame showed less variability than those conducted with multiple specimens per creep frame.
- The variability in the creep index when a multi-specimen setup is used can be reduced by casting and pre-cracking more specimens than strictly needed for the creep tests and testing the most similar ones.

- In notched beams, if the CMOD is not measured with a clip transducer fitted to the crack mouth, shrinkage between measuring points should be considered. Otherwise, the shrinkage deformations may affect the assessment of the delayed displacements due to creep.
- Although the limited number of specimens in direct tension hindered the assessment of the influence of creep index, concrete strength or testing procedure, the creep behaviour shows a clear dependence on the fibre type.
- In the case of low fibre dosages, the flexure creep tests procedure may lead to relatively high scatter as observed in the quasi-static characterisation of the post-cracking residual stress. A minimum structural performance can be recommended to mitigate such variability.
- The construction of creep frames where the specimens are hold independently by means of steel plates is an effective way to mitigate the impact of failure of one specimen on the others in the same column and not affect or case the interruption of the creep test.
- The use of an upper lever arm is the most appropriate approach for these tests given its simplicity. The upper lever arm configuration is preferred since the load is applied directly over the upper specimen, reducing time for the loading procedure and the influence of friction and of other elements.
- The support or loading rollers placed between specimens and in the contact with the frame influence the freedom of movement of the specimens and, consequently, the displacements measured in the creep test. Therefore, the detailed description and the standardisation of the boundary conditions are needed to ensure similar conditions and obtain comparable results.
- The time in which load is applied becomes a key parameter since influence directly into the creep coefficient because defines the instantaneous deformations. A restriction in time should be given in a future standard procedure in order to obtain comparable results.
- The influence of secondary parameters such as fibre count and distribution, humidity, temperature and minor differences in the testing procedure should be the subject of additional studies.
- The creep results are quite sensitive to the environmental conditions such as temperature and, mainly, humidity which is difficult to control and keep constant during the long duration of creep tests. In a reference creep test methodology, the temperature and relative humidity ranges should be defined and the evolution of climate conditions and all climatic issues during the creep test duration should be reported.
- In order to guarantee "controlled" climatic conditions, hygrometric and temperature equipment are required in order to keep the temperature and humidity in a certain defined threshold. On the contrary, if creep tests are performed in climatic or insulated room without any hygrometric equipment or temperature control, the climatic conditions should be defined only as "restricted".
- Although specimens may be wrapped to reduce exchanges with the environment and mitigate the difficulty of controlling the humidity, it is recommended to measure the shrinkage in pre-cracked specimens without load.

- The creep coefficient defined in the RRT is analogous to the creep coefficients traditionally used for structural evaluation of creep in codes. The creep coefficient referred to creep period and the creep coefficient referred to "origin" should be assessed, since different creep coefficient values are obtained.
- The time in which the load is applied at the beginning of the creep test is a significant issue that present a great variability among the participants. Although no clear influence on the general creep curve over time was assessed, this parameter affects the instantaneous deformations, which has a direct impact on the calculation of the creep coefficient and SyFRC mixes are clearly sensitive to this parameter. A maximum duration should be established in a future standard to ensure comparable results.
- Creep coefficient referred to the creep period from SyFRC mixes were higher and more scattered than for SFRC mixes. These differences depend on the different physical behaviour of the two composite materials, as well as differences in the procedures and different criteria of elastic deformation definition.
- Regarding the different $CMOD_{ci}$ references analysed it can be concluded that the consideration of the start of the creep stage at $t_0 + 30'$ led to more homogeneous creep coefficients since the sensitivity to the elastic deformation reduces when the initial time increase.
- SyFRC mixes are more sensitive to the definition of elastic deformation than SFRC due to the high creep rates that synthetic fibres exhibit in the early stages of creep tests. It is crucial to reach a consensus about the definition of the instantaneous deformation to ensure more robust and reproducible results.
- It was observed for both SyFRC and SFRC mixes that the total variability is decomposed into between- and within- laboratory components: the between-laboratories variance component is higher that the within- event component. This means that the differences between mean values of the creep coefficients for the different laboratories are higher that the differences observed among the specimens tested by each laboratory. Therefore, the testing procedures adopted were revealed as a significant factor.
- High correlations to f_R and I_c were found by SyFRC mixes since creep coefficients tends to decrease as f_R increases and to increase with I_c. Notwithstanding, there also exists a correlation to other factors such as the type of supports or the load configuration. In the case of the support type, the presence of supports not allowing horizontal movements is correlated to higher creep coefficient values, whereas in the case of the loading configuration the 3PBT configuration leads to higher values.
- In the case of the analysis of SFRC creep coefficients, only the factors concerning the type of supports were revealed as statistically significant.
- The COR evolution is difficult to evaluate in short time lapses using discrete measurements given the high influence of the testing procedure and environmental conditions. More consistent COR assessment can be obtained using a continuous CMOD trend line.

- The consideration of the shrinkage is essential in the COR calculation, particularly at earlier times. By contrast, in most cases, the influence of the shrinkage may be disregarded if COR is evaluated for time lapses after 90 days.
- The COR may be considered a less robust parameter than the creep coefficient given the COR sensitivity to variations in test procedure and environmental conditions. Additional studies are needed to evaluate if this observation also stands when COR are assessed for CMOD trendlines.
- The fibre counting process must be better defined in order to improve the ulterior analyses and avoid double counting, in particular for SyFRCs.

Appendix A
Specimens Production Index

Aitor Llano-Torre and Pedro Serna

The day after production, the specimens were demoulded, marked with the batch reference and numbered with an identification number (ID) according to the casting order to identify potential differences induced by the production order. Table A.1 presents the corresponding numbers of specimens for the different FRC batches.

Table A.2 indexes the produced specimens arranged by the specimen ID. This classification provides information about the destination of each specimen and enables to track if the production order may influence the mechanical performance. Additional information such as shape and dimensions is also provided.

Table A.3 indexes the different specimens arranged by the laboratory destination to classify the check how many specimens were delivered to each participant. Additional information such as shape and dimensions is also provided.

Table A.1 Identification of the specimens from different batches

Mix	Batch	ID number	N° Specimens
SyFRC	M-B1	For specimen numbers from 1 to 112	112
	M-B2	For specimen numbers from 113 to 221	109
SFRC	S-B1	For specimen numbers from 222 to 328	107
	S-B2	For specimen numbers from 329 to 404	76
	S-B0	For specimen numbers from 405 to 458	54

A. Llano-Torre · P. Serna
Institute of Concrete Science and Technology ICITECH, Universitat Politècnica de València (UPV), Valencia, Spain
e-mail: aillator@upv.es

P. Serna
e-mail: pserna@cst.upv.es

© RILEM 2021
A. Llano-Torre and P. Serna (eds.), *Round-Robin Test on Creep Behaviour in Cracked Sections of FRC: Experimental Program, Results and Database Analysis*, RILEM State-of-the-Art Reports 34, https://doi.org/10.1007/978-3-030-72736-9

Table A.2 Specimens index arranged by FRC batch and ID number

Specimen ID	Shape	Dimensions (mm)	Destination	Amount
M-B1-001 to M-B1-020	Cylindrical	Ø150 × 300	LAB-01	20
M-B1-021 to M-B1-022	Cylindrical	Ø150 × 300	LAB-05	2
M-B1-023 to M-B1-030	Cylindrical	Ø150 × 300	LAB-11	8
M-B1-031 to M-B1-035	Prismatic 3	150 × 150 × 700	LAB-09	5
M-B1-036 to M-B1-037	Square Panel	600 × 600 × 100	LAB-12	2
M-B1-038 to M-B1-042	Square Panel	600 × 600 × 100	LAB-15	5
M-B1-043 to M-B1-044	Round Panel	Ø800 × 75	LAB-17	2
M-B1-045 to M-B1-054	Prismatic 1	150 × 150 × 600	LAB-01	10
M-B1-055 to M-B1-059	Prismatic 1	150 × 150 × 600	LAB-02	5
M-B1-060 to M-B1-064	Prismatic 1	150 × 150 × 600	LAB-03	5
M-B1-065 to M-B1-074	Prismatic 1	150 × 150 × 600	LAB-04	10
M-B1-075 to M-B1-079	Prismatic 1	150 × 150 × 600	LAB-05	5
M-B1-080 to M-B1-084	Prismatic 1	150 × 150 × 600	LAB-06	5
M-B1-085 to M-B1-092	Prismatic 1	150 × 150 × 600	LAB-11	8
M-B1-093 to M-B1-097	Prismatic 1	150 × 150 × 600	LAB-12	5
M-B1-098 to M-B1-107	Prismatic 1	150 × 150 × 600	LAB-01	10
M-B1-108 to M-B1-111	Prismatic 2	100 × 100 × 500	LAB-03	4
M-B1-112	Prismatic 2	100 × 100 × 500	LAB-08	1
M-B2-113 to M-B2-132	Cylindrical	Ø150 × 300	LAB-01	20
M-B2-133 to M-B2-140	Cylindrical	Ø150 × 300	LAB-11	8
M-B2-141	Cylindrical	Ø150 × 300	LAB-01	1
M-B2-142to M-B2-146	Square Panel	600 × 600 × 100	LAB-15	5
M-B2-147to M-B2-150	Square Panel	600 × 600 × 100	LAB-18	4
M-B2-151to M-B2-152	Round Panel	Ø800 × 75	LAB-17	2
M-B2-153 to M-B2-162	Prismatic 1	150 × 150 × 600	LAB-01	10
M-B2-163 to M-B2-167	Prismatic 1	150 × 150 × 600	LAB-02	5
M-B2-168 to M-B2-176	Prismatic 1	150 × 150 × 600	LAB-04	9
M-B2-177 to M-B2-181	Prismatic 1	150 × 150 × 600	LAB-05	5
M-B2-182 to M-B2-186	Prismatic 1	150 × 150 × 600	LAB-07	5
M-B2-187 to M-B2-190	Prismatic 1	150 × 150 × 600	LAB-08	4
M-B2-191 to M-B2-194	Prismatic 1	150 × 150 × 600	LAB-10	4
M-B2-195 to M-B2-196	Prismatic 1	150 × 150 × 600	LAB-11	2
M-B2-197	Prismatic 1	150 × 150 × 600	LAB-14	1
M-B2-198 to M-B2-202	Prismatic 1	150 × 150 × 600	LAB-19	5
M-B2-203 to M-B2-213	Prismatic 1	150 × 150 × 600	LAB-01	11
M-B2-214 to M-B2-221	Prismatic 2	100 × 100 × 500	LAB-16	8

(continued)

Table A.2 (continued)

Specimen ID	Shape	Dimensions (mm)	Destination	Amount
S-B1-222 to S-B1-241	Cylindrical	Ø150 × 300	LAB-01	20
S-B1-242 to S-B1-243	Cylindrical	Ø150 × 300	LAB-05	2
S-B1-244 to S-B1-251	Cylindrical	Ø150 × 300	LAB-11	8
S-B1-252 to S-B1-256	Prismatic 3	150 × 150 × 700	LAB-09	5
S-B1-257 to S-B1-259	Prismatic 3	150 × 150 × 700	LAB-13	3
S-B1-260 to S-B1-261	Square Panel	600 × 600 × 100	LAB-12	2
S-B1-262 to S-B1-266	Square Panel	600 × 600 × 100	LAB-15	5
S-B1-267 to S-B1-276	Prismatic 1	150 × 150 × 600	LAB-01	10
S-B1-277 to S-B1-281	Prismatic 1	150 × 150 × 600	LAB-02	5
S-B1-282 to S-B1-286	Prismatic 1	150 × 150 × 600	LAB-03	5
S-B1-287 to S-B1-294	Prismatic 1	150 × 150 × 600	LAB-04	8
S-B1-295 to S-B1-299	Prismatic 1	150 × 150 × 600	LAB-05	5
S-B1-300 to S-B1-304	Prismatic 1	150 × 150 × 600	LAB-06	5
S-B1-305 to S-B1-312	Prismatic 1	150 × 150 × 600	LAB-11	8
S-B1-313 to S-B1-317	Prismatic 1	150 × 150 × 600	LAB-12	5
S-B1-318 to S-B1-327	Prismatic 1	150 × 150 × 600	LAB-01	10
S-B1-328	Prismatic 2	100 × 100 × 500	LAB-08	1
S-B2-329 to S-B2-342	Cylindrical	Ø150 × 300	LAB-01	14
S-B2-343 to S-B2-348	Cylindrical	Ø150 × 300	LAB-11	6
S-B2-349 to S-B2-353	Square Panel	600 × 600 × 100	LAB-15	5
S-B2-354 to S-B2-357	Square Panel	600 × 600 × 100	LAB-18	4
S-B2-358 to S B2-367	Prismatic 1	150 × 150 × 600	LAB-01	10
S-B2-368 to S-B2-383	Prismatic 1	150 × 150 × 600	LAB-02	5
S-B2-373 to S-B2-381	Prismatic 1	150 × 150 × 600	LAB-04	9
S-B2-382	Prismatic 1	150 × 150 × 600	LAB-11	2
S-B2-384	Prismatic 1	150 × 150 × 600	LAB-14	1
S-B2-385 to S-B2-396	Prismatic 1	150 × 150 × 600	LAB-01	12
S-B2-397 to S-B2-404	Prismatic 2	100 × 100 × 500	LAB-16	8
S-B0-405to S-B0-410	Cylindrical	Ø150 × 300	LAB-01	6
S-B0-411 to S-B0-413	Square Panel	600 × 600 × 100	LAB-01	3
S-B0-414 to S-B0-428	Prismatic 1	150 × 150 × 600	LAB-01	15
S-B0-429 to S-B0-433	Prismatic 1	150 × 150 × 600	LAB-05	5
S-B0-434 to S-B0-438	Prismatic 1	150 × 150 × 600	LAB-06	5
S-B0-439 to S-B0-443	Prismatic 1	150 × 150 × 600	LAB-07	5
S-B0-444 to S-B0-449	Prismatic 1	150 × 150 × 600	LAB-08	6
S-B0-450 to S-B0-453	Prismatic 1	150 × 150 × 600	LAB-10	4
S-B0-454 to S-B0-458	Prismatic 1	150 × 150 × 600	LAB-19	5

Table A.3 Specimens index arranged by destination

Destination	Specimen ID	Shape	Dimensions (mm)	Amount
LAB-01	M-B1-001 to M-B1-020	Cylindrical	$\varnothing 150 \times 300$	20
	M-B1-045 to M-B1-054	Prismatic 1	$150 \times 150 \times 600$	10
	M-B1-098 to M-B1-107	Prismatic 1	$150 \times 150 \times 600$	10
	M-B2-113 to M-B2-132	Cylindrical	$\varnothing 150 \times 300$	20
	M-B2-141	Cylindrical	$\varnothing 150 \times 300$	1
	M-B2-153 to M-B2-162	Prismatic 1	$150 \times 150 \times 600$	10
	M-B2-203 to M-B2-213	Prismatic 1	$150 \times 150 \times 600$	11
	S-B1-222 to S-B1-241	Cylindrical	$\varnothing 150 \times 300$	20
	S-B1-267 to S-B1-276	Prismatic 1	$150 \times 150 \times 600$	10
	S-B1-318 to S-B1-327	Prismatic 1	$150 \times 150 \times 600$	10
	S-B2-329 to S-B2-342	Cylindrical	$\varnothing 150 \times 300$	14
	S-B2-358 to S-B2-367	Prismatic 1	$150 \times 150 \times 600$	10
	S-B2-385 to S-B2-396	Prismatic 1	$150 \times 150 \times 600$	12
	S-B0-405 to S-B0-410	Cylindrical	$\varnothing 150 \times 300$	6
	S-B0-411 to S-B0-413	Square Panel	$600 \times 600 \times 100$	3
	S-B0-414 to S-B0-428	Prismatic 1	$150 \times 150 \times 600$	15
LAB-02	M-B1-055 to M-B1-059	Prismatic 1	$150 \times 150 \times 600$	5
	M-B2-163 to M-B2-167	Prismatic 1	$150 \times 150 \times 600$	5
	S-B1-277 to S-B1-281	Prismatic 1	$150 \times 150 \times 600$	5
	S-B2-368 to S-B2-372	Prismatic 1	$150 \times 150 \times 600$	5
LAB-03	M-B1-060 to M-B1-064	Prismatic 1	$150 \times 150 \times 600$	5
	S-B1-282 to S-B1-286	Prismatic 1	$150 \times 150 \times 600$	5
	M-B1-108 to M-B1-111	Prismatic 2	$100 \times 100 \times 500$	4
LAB-04	M-B1-065 to M-B1-074	Prismatic 1	$150 \times 150 \times 600$	10
	M-B2-168 to M-B2-176	Prismatic 1	$150 \times 150 \times 600$	9
	S-B1-287 to S-B1-294	Prismatic 1	$150 \times 150 \times 600$	8
	S-B2-373 to S-B2-381	Prismatic 1	$150 \times 150 \times 600$	9
LAB-05	M-B1-021 to M-B1-022	Cylindrical	$\varnothing 150 \times 300$	2
	M-B1-075 to M-B1-079	Prismatic 1	$150 \times 150 \times 600$	5
	M-B2-177 to M-B2-181	Prismatic 1	$150 \times 150 \times 600$	5
	S-B1-242 to S-B1-243	Cylindrical	$\varnothing 150 \times 300$	2
	S-B1-295 to S-B1-299	Prismatic 1	$150 \times 150 \times 600$	5
	S-B0-429 to S-B0-433	Prismatic 1	$150 \times 150 \times 600$	5
LAB-06	M-B1-080 to M-B1-084	Prismatic 1	$150 \times 150 \times 600$	5
	S-B1-300 to S-B1-304	Prismatic 1	$150 \times 150 \times 600$	5
	S-B0-434 to S-B0-438	Prismatic 1	$150 \times 150 \times 600$	5

(continued)

Table A.3 (continued)

Destination	Specimen ID	Shape	Dimensions (mm)	Amount
LAB-07	M-B2-182 to M-B2-186	Prismatic 1	$150 \times 150 \times 600$	5
	S-B0-439 to S-B0-443	Prismatic 1	$150 \times 150 \times 600$	5
LAB-08	M-B1-112	Prismatic 2	$100 \times 100 \times 500$	1
	M-B2-187 to M-B2-190	Prismatic 1	$150 \times 150 \times 600$	4
	S-B1-328	Prismatic 2	$100 \times 100 \times 500$	1
	S-B0-444 to S-B0-449	Prismatic 1	$150 \times 150 \times 600$	6
LAB-09	M-B1-031 to M-B1-035	Prismatic 3	$150 \times 150 \times 700$	5
	S-B1-252 to S-B1-256	Prismatic 3	$150 \times 150 \times 700$	5
LAB-10	M-B2-191 to M-B2-194	Prismatic 1	$150 \times 150 \times 600$	4
	S-B0-450 to S-B0-453	Prismatic 1	$150 \times 150 \times 600$	4
LAB-11	M-B1-023 to M-B1-030	Cylindrical	$\varnothing 150 \times 300$	8
	M-B1-085 to M-B1-092	Prismatic 1	$150 \times 150 \times 600$	8
	M-B2-133 to M-B2-140	Cylindrical	$\varnothing 150 \times 300$	8
	M-B2-195 to M-B2-196	Prismatic 1	$150 \times 150 \times 600$	2
	S-B1-244 to S-B1-251	Cylindrical	$\varnothing 150 \times 300$	8
	S-B1-305 to S-B1-312	Prismatic 1	$150 \times 150 \times 600$	8
	S-B2-343 to S-B2-348	Cylindrical	$\varnothing 150 \times 300$	6
	S-B2-382 to S-B2-383	Prismatic 1	$150 \times 150 \times 600$	2
LAB-12	M-B1-036 to M-B1-037	Square Panel	$600 \times 600 \times 100$	2
	M-B1-093 to M-B1-097	Prismatic 1	$150 \times 150 \times 600$	5
	S-B1-260 to S-B1-261	Square Panel	$600 \times 600 \times 100$	2
	S-B1-313 to S-B1-317	Prismatic 1	$150 \times 150 \times 600$	5
LAB-13	S-B1-257 to S-B1-259	Prismatic 3	$150 \times 150 \times 700$	3
LAB-14	M-B2-197	Prismatic 1	$150 \times 150 \times 600$	1
	S-B2-384	Prismatic 1	$150 \times 150 \times 600$	1
LAB-15	M-B1-038 to M-B1-042	Square Panel	$600 \times 600 \times 100$	5
	M-B2-142 to M-B2-146	Square Panel	$600 \times 600 \times 100$	5
	S-B1-262 to S-B1-266	Square Panel	$600 \times 600 \times 100$	5
	S-B2-349 to S-B2-353	Square Panel	$600 \times 600 \times 100$	5
LAB-16	M-B2-214 to M-B2-221	Prismatic 2	$100 \times 100 \times 500$	8
	S-B2-397 to S-B2-404	Prismatic 2	$100 \times 100 \times 500$	8
LAB-17	M-B1-043 to M-B1-044	Round Panel	$\varnothing 800 \times 75$	2
	M-B2-151 to M-B2-152	Round Panel	$\varnothing 800 \times 75$	2
LAB-18	M-B2-147 to M-B2-150	Square Panel	$600 \times 600 \times 100$	4
	S-B2-354 to S-B2-357	Square Panel	$600 \times 600 \times 100$	4
LAB-19	M-B2-198 to M-B2-202	Prismatic 1	$150 \times 150 \times 600$	5
	S-B0-454 to S-B0-458	Prismatic 1	$150 \times 150 \times 600$	5

Appendix B
Characterization Test

Aitor Llano-Torre and Pedro Serna

During the RRT execution, characterization tests were performed to the different FRC batches along the duration of creep test at different time lapses: 7 days, 28 days, T1 (start of creep test), T2 (6 months of creep test) and T3 (end of creep test). Full information about the results of characterization test, compressive strength and flexural strength test, are available in this Appendix.

Batch M-B1 Characterization Test Results

Batch M-B1 Compressive Strength Test Results

A total of 15 cylindrical specimens were tested to failure in compression. The compressive strength test results are shown in Table B.1.

A. Llano-Torre · P. Serna
Institute of Concrete Science and Technology ICITECH, Universitat Politècnica de València (UPV), Valencia, Spain
e-mail: aillator@upv.es

P. Serna
e-mail: pserna@cst.upv.es

© RILEM 2021
A. Llano-Torre and P. Serna (eds.), *Round-Robin Test on Creep Behaviour in Cracked Sections of FRC: Experimental Program, Results and Database Analysis*, RILEM State-of-the-Art Reports 34, https://doi.org/10.1007/978-3-030-72736-9

Table B.1 Compressive strength characterization results from batch M-B1

Specimen	Compressive strength f_c (MPa)				
	7 days	28 days	T1	T2	T3
M-B1-001	30.2	–	–	–	–
M-B1-002	30.8	–	–	–	–
M-B1-003	30.6	–	–	–	–
M-B1-004	–	33.8	–	–	–
M-B1-005	–	33.6	–	–	–
M-B1-006	–	34.4	–	–	–
M-B1-007	–	–	38.3	–	–
M-B1-008	–	–	37.5	–	–
M-B1-009	–	–	37.8	–	–
M-B1-010	–	–	–	39.95	–
M-B1-011	–	–	–	40.35	–
M-B1-012	–	–	–	40.03	–
M-B1-013	–	–	–	–	43.5
M-B1-014	–	–	–	–	41.3
M-B1-015	–	–	–	–	41.7
Mean	**30.53**	**33.93**	**37.85**	**40.11**	**42.18**
CV (%)	1.0	1.2	1.1	0.5	2.8

Batch M-B1 Flexural Test Results

Flexural characterization tests were performed at different ages in 15 prismatic specimens from batch M-B1. The results obtained at different ages are summarised in Figs. B.1, B.2, B.3, B.4 and B.5 at 7 days, 28 days, T1, T2 and T3 respectively.

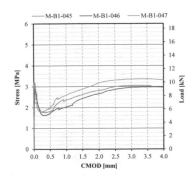

Specimen	f_L	$f_{R,1}$	$f_{R,2}$	$f_{R,3}$	$f_{R,4}$
M-B1-045	3.19	1.83	2.63	2.99	3.05
M-B1-046	2.97	1.73	2.44	2.87	2.97
M-B1-047	3.20	2.03	2.88	3.31	3.36
Mean	**3.12**	**1.86**	**2.65**	**3.06**	**3.13**
CV [%]	4.1	8.1	8.4	7.3	6.6

Fig. B.1 M-B1 flexural residual strength characterization: test results at 7 days

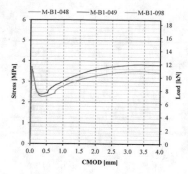

Specimen	f_L	$f_{R,1}$	$f_{R,2}$	$f_{R,3}$	$f_{R,4}$
M-B1-048	--	--	--	--	--
M-B1-049	3.58	2.45	3.35	3.72	3.82
M-B1-098	3.72	2.30	3.08	3.42	3.50
Mean	**3.65**	**2.37**	**3.21**	**3.57**	**3.66**
CV [%]	2.8	4.7	5.9	6.1	6.2

Fig. B.2 M-B1 flexural residual strength characterization: test results at 28 days

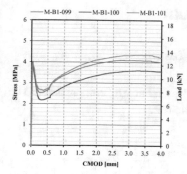

Specimen	f_L	$f_{R,1}$	$f_{R,2}$	$f_{R,3}$	$f_{R,4}$
M-B1-099	4.02	2.65	3.70	4.06	4.07
M-B1-100	3.97	2.26	3.12	3.51	3.60
M-B1-101	3.96	2.73	3.84	4.25	4.34
Mean	**3.98**	**2.55**	**3.56**	**3.94**	**4.00**
CV [%]	0.9	9.8	10.7	9.7	9.4

Fig. B.3 M-B1 flexural residual strength characterization: test results at T1

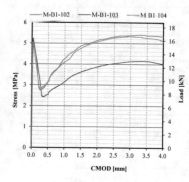

Specimen	f_L	$f_{R,1}$	$f_{R,2}$	$f_{R,3}$	$f_{R,4}$
M-B1-102	5.01	3.13	4.86	5.31	5.38
M-B1-103	5.23	2.54	3.62	4.04	4.14
M-B1-104	5.10	3.16	4.76	5.26	5.26
Mean	**5.11**	**2.95**	**4.41**	**4.87**	**4.93**
CV [%]	2.2	11.9	15.6	14.8	13.9

Fig. B.4 M-B1 flexural residual strength characterization: test results at T2

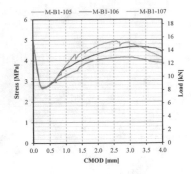

Specimen	f_L	$f_{R,1}$	$f_{R,2}$	$f_{R,3}$	$f_{R,4}$
M-B1-105	4.74	2.82	3.74	4.13	4.03
M-B1-106	4.61	2.82	4.02	4.54	4.64
M-B1-107	4.93	2.87	4.39	4.93	4.53
Mean	**4.76**	**2.84**	**4.05**	**4.53**	**4.40**
CV [%]	3.4	0.9	8.0	8.8	7.3

Fig. B.5 M-B1 flexural residual strength characterization: test results at T3

Batch M-B2 Characterization Test Results

Batch M-B2 Compressive Strength Test Results

A total of 15 cylindrical specimens were tested to failure in compression. The compressive strength test results are shown in Table B.2.

Table B.2 Compressive strength characterization results from batch M-B2

Specimen	Compressive strength f_c (MPa)				
	7 days	28 days	T1	T2	T3
M-B2-113	30.0	–	–	–	–
M-B2-114	31.3	–	–	–	–
M-B2-115	32.4	–	–	–	–
M-B2-116	–	37.4	–	–	–
M-B2-117	–	37.7	–	–	–
M-B2-118	–	37.6	–	–	–
M-B2-119	–	–	42.6	–	–
M-B2-120	–	–	42.9	–	–
M-B2-121	–	–	42.0	–	–
M-B2-122	–	–	–	39.19	–
M-B2-123	–	–	–	36.12	–
M-B2-124	–	–	–	41.15	–
M-B2-125	–	–	–	–	46.1
M-B2-126	–	–	–	–	45.1
M-B2-127	–	–	–	–	41.0
Mean	**31.23**	**37.57**	**42.46**	**38.82**	**44.06**
CV (%)	3.8	0.4	1.1	6.5	6.1

Batch M-B2 Flexural Test Results

Flexural characterization tests were performed in 15 prismatic specimens from batch M-B2. The results obtained at different ages are summarised in Figs. B.6, B.7, B.8,

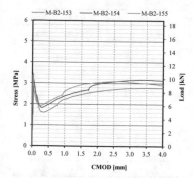

Specimen	f_L	$f_{R,1}$	$f_{R,2}$	$f_{R,3}$	$f_{R,4}$
M-B2-153	3.46	1.70	2.47	2.71	2.79
M-B2-154	3.06	1.98	2.63	3.08	3.17
M-B2-155	3.34	2.12	2.93	3.09	3.00
Mean	**3.29**	**1.93**	**2.68**	**2.96**	**2.98**
CV [%]	6.3	11.0	8.9	7.2	6.4

Fig. B.6 M-B2 flexural residual strength characterization: test results at 7 days

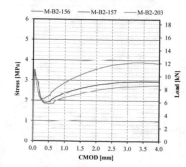

Specimen	f_L	$f_{R,1}$	$f_{R,2}$	$f_{R,3}$	$f_{R,4}$
M-B2-156	3.32	2.22	3.26	3.75	3.88
M-B2-157	3.52	1.93	2.61	2.87	2.94
M-B2-203	3.52	1.83	2.34	2.63	2.72
Mean	**3.45**	**1.99**	**2.74**	**3.08**	**3.18**
CV [%]	3.3	10.1	17.2	19.2	19.4

Fig. B.7 M-B2 flexural residual strength characterization: test results at 28 days

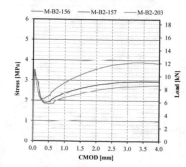

Specimen	f_L	$f_{R,1}$	$f_{R,2}$	$f_{R,3}$	$f_{R,4}$
M-B2-204	3.92	2.06	2.61	3.18	3.20
M-B2-205	3.90	1.53	2.03	2.34	2.50
M-B2-206	3.95	2.15	3.03	3.41	3.47
Mean	**3.92**	**1.92**	**2.56**	**2.97**	**3.06**
CV [%]	0.7	17.5	19.6	19.0	16.4

Fig. B.8 M-B2 flexural residual strength characterization: test results at T1

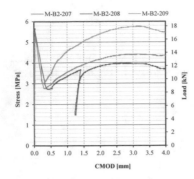

Specimen	f_L	$f_{R,1}$	$f_{R,2}$	$f_{R,3}$	$f_{R,4}$
M-B2-207	5.44	2.95	3.90	4.34	4.37
M-B2-208	5.67	2.76	3.49	3.93	3.88
M-B2-209	5.26	3.47	5.02	5.65	5.68
Mean	**5.45**	**3.06**	**4.14**	**4.64**	**4.64**
CV [%]	3.7	12.0	19.1	19.3	20.1

Fig. B.9 M-B2 flexural residual strength characterization: test results at T2

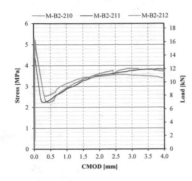

Specimen	f_L	$f_{R,1}$	$f_{R,2}$	$f_{R,3}$	$f_{R,4}$
M-B2-210	5.21	2.26	3.35	3.69	3.82
M-B2-211	4.36	2.37	3.28	3.64	3.83
M-B2-212	5.32	2.63	3.43	3.57	3.50
Mean	**4.96**	**2.42**	**3.35**	**3.63**	**3.72**
CV [%]	10.6	7.9	2.3	1.6	5.0

Fig. B.10 M-B2 flexural residual strength characterization: test results at T3

B.9 and B.10 at 7 days, 28 days, T1, T2 and T3 respectively.

Batch S-B1 Characterization Test Results

Batch S-B1 Compressive Strength Test Results

A total of 15 cylindrical specimens were tested to failure in compression. The compressive strength test results are shown in Table B.3.

Batch S-B1 Flexural Test Results

Flexural characterization tests were performed in 15 prismatic specimens from batch S-B1. The results obtained at different ages are summarised in Figs. B.11, B.12, B.13,

Table B.3 S-B1 batch characterization: compressive strength test results

Specimen	Compressive strength f_c (MPa)				
	7 days	28 days	T1	T2	T3
S-B1-222	27.5	–	–	–	–
S-B1-223	27.2	–	–	–	–
S-B1-224	27.3	–	–	–	–
S-B1-225	–	29.5	–	–	–
S-B1-226	–	31.6	–	–	–
S-B1-227	–	31.3	–	–	–
S-B1-228	–	–	35.3	–	–
S-B1-229	–	–	34.3	–	–
S-B1-230	–	–	34.6	–	–
S-B1-231	–	–	–	35.24	–
S-B1-232	–	–	–	35.22	–
S-B1-233	–	–	–	32.89	–
S-B1-234	–	–	–	–	34.8
S-B1-235	–	–	–	–	37.6
S-B1-236	–	–	–	–	34.2
Mean	**27.33**	**30.80**	**34.73**	**34.45**	**35.53**
CV (%)	0.6	3.7	1.4	3.9	5.1

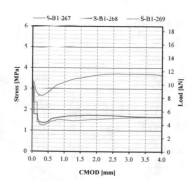

Specimen	f_L	$f_{R,1}$	$f_{R,2}$	$f_{R,3}$	$f_{R,4}$
S-B1-267	3.36	2.85	3.52	3.71	3.71
S-B1-268	3.10	1.44	1.72	1.72	1.67
S-B1-269	2.99	1.35	1.49	1.58	1.64
Mean	**3.15**	**1.88**	**2.24**	**2.34**	**2.34**
CV [%]	6.0	44.7	49.4	51.0	50.7

Fig. B.11 S-B1 flexural residual strength characterization: test results at 7 days

B.14 and B.15 at 7 days, 28 days, T1, T2 and T3 respectively.

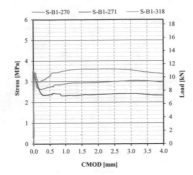

Specimen	f_L	$f_{R,1}$	$f_{R,2}$	$f_{R,3}$	$f_{R,4}$
S-B1-270	3.55	2.74	2.93	2.99	3.02
S-B1-271	3.32	2.37	2.35	2.40	2.37
S-B1-318	3.44	3.28	3.59	3.60	3.45
Mean	**3.44**	**2.80**	**2.96**	**3.00**	**2.94**
CV [%]	3.5	16.3	21.1	20.0	18.5

Fig. B.12 S-B1 flexural residual strength characterization: test results at 28 days

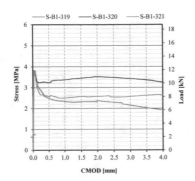

Specimen	f_L	$f_{R,1}$	$f_{R,2}$	$f_{R,3}$	$f_{R,4}$
S-B1-319	3.82	2.49	2.31	2.28	2.03
S-B1-320	3.82	3.24	3.44	3.48	3.35
S-B1-321	3.83	2.58	2.57	2.54	2.63
Mean	**3.82**	**2.77**	**2.78**	**2.77**	**2.67**
CV [%]	0.1	14.9	21.3	22.7	24.9

Fig. B.13 S-B1 flexural residual strength characterization: test results at T1

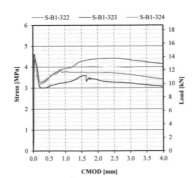

Specimen	f_L	$f_{R,1}$	$f_{R,2}$	$f_{R,3}$	$f_{R,4}$
S-B1-322	4.58	3.50	4.28	4.41	4.23
S-B1-323	4.61	3.06	3.56	3.27	3.14
S-B1-324	4.50	3.48	3.74	3.71	3.53
Mean	**4.56**	**3.35**	**3.86**	**3.79**	**3.63**
CV [%]	1.2	7.4	9.6	15.1	15.2

Fig. B.14 S-B1 flexural residual strength characterization: test results at T2

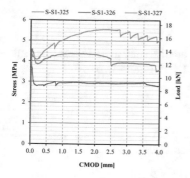

Specimen	f_L	$f_{R,1}$	$f_{R,2}$	$f_{R,3}$	$f_{R,4}$
S-B1-325	4.58	4.11	4.36	3.81	3.88
S-B1-326	3.92	2.87	2.97	2.94	2.96
S-B1-327	4.31	4.58	5.32	5.54	5.19
Mean	**4.27**	**3.85**	**4.21**	**4.10**	**4.01**
CV [%]	7.8	22.9	28.0	32.4	27.9

Fig. B.15 S-B1 flexural residual strength characterization: test results at T3

Batch S-B2 Characterization Test Results

Batch S-B2 Compressive Strength Test Results

A total of 10 cylindrical specimens were tested to failure in compression. The compressive strength test results are shown in Table B.4.

Table B.4 S-B2 batch characterization: compressive strength test results

Specimen	Compressive strength f_c (MPa)				
	7 days	28 days	T1	T2	T3
S B2-329	30.5	–	–	–	–
S-B2-330	31.3	–	–	–	–
S-B2-331	30.5	–	–	–	–
S-B2-332	–	36.7	–	–	–
S-B2-333	–	36.3	–	–	–
S-B2-334	–	35.5	–	–	–
S-B2-335	–	–	38.7	–	–
S-B2-336	–	–	37.1	–	–
S-B2-337	–	–	–	38.88	
S-B2-338	–	–	–	38.29	
Mean	**30.77**	**36.17**	**37.93**	**38.59**	–
CV (%)	1.5	1.7	3.0	1.1	–

Batch S-B2 Flexural Test Results

Flexural characterization tests were performed at different ages in 15 prismatic specimens from batch S-B2 (Figs. B.16, B.17, B.18, B.19 and B.20).

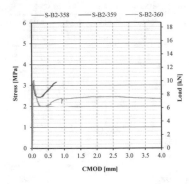

Specimen	f_L	$f_{R,1}$	$f_{R,2}$	$f_{R,3}$	$f_{R,4}$
S-B2-358	--	--	--	--	--
S-B2-359	3.08	2.80	--	--	--
S-B2-360	3.23	2.05	2.42	2.46	2.38
Mean	**3.16**	**2.42**	**2.42**	**2.46**	**2.38**
CV [%]	3.4	21.6	--	--	--

Fig. B.16 S-B2 flexural residual strength characterization: test results at 7 days

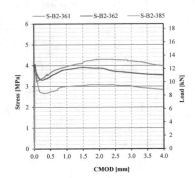

Specimen	f_L	$f_{R,1}$	$f_{R,2}$	$f_{R,3}$	$f_{R,4}$
S-B2-361	3.94	3.64	4.13	4.29	4.11
S-B2-362	4.06	3.46	3.90	3.72	3.58
S-B2-385	3.71	2.74	3.04	3.06	2.91
Mean	**3.90**	**3.28**	**3.69**	**3.69**	**3.53**
CV [%]	4.5	14.6	15.6	16.7	17.0

Fig. B.17 S-B2 flexural residual strength characterization: test results at 28 days

Specimen	f_L	$f_{R,1}$	$f_{R,2}$	$f_{R,3}$	$f_{R,4}$
S-B2-386	3.52	3.07	3.49	3.43	3.29
S-B2-387	4.23	3.48	4.07	4.14	4.05
S-B2-388	3.97	4.81	5.59	5.56	5.37
Mean	**3.90**	**3.79**	**4.38**	**4.38**	**4.23**
CV [%]	9.3	24.1	24.8	24.8	24.8

Fig. B.18 S-B2 flexural residual strength characterization: test results at T1

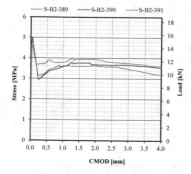

Specimen	f_L	$f_{R,1}$	$f_{R,2}$	$f_{R,3}$	$f_{R,4}$
S-B2-389	4.98	3.33	3.60	3.58	3.32
S-B2-390	4.93	3.26	3.75	3.66	3.60
S-B2-391	4.61	3.80	3.95	3.82	3.71
Mean	**4.84**	**3.47**	**3.76**	**3.69**	**3.54**
CV [%]	4.2	8.4	4.7	3.2	5.7

Fig. B.19 S-B2 flexural residual strength characterization: test results at T2

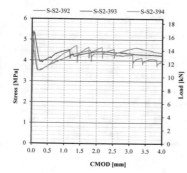

Specimen	f_L	$f_{R,1}$	$f_{R,2}$	$f_{R,3}$	$f_{R,4}$
S-B2-392	4.48	3.74	4.36	4.27	4.00
S-B2-393	5.06	4.09	4.28	4.38	4.28
S-B2-394	5.42	3.76	4.39	4.53	4.52
Mean	**4.99**	**3.86**	**4.35**	**4.39**	**4.27**
CV [%]	9.5	5.0	1.3	3.0	6.1

Fig. B.20 S-B2 flexural residual strength characterization: test results at T3

Batch S-B0 Characterization Test Results

Batch S-B0 Compressive Strength Test Results

A total of 6 cylindrical specimens were tested to failure in compression. The compressive strength results are shown in Table B.5.

Batch S-B0 Flexural Test Results

Flexural characterization tests were performed in 12 prismatic specimens from batch S-B0. The results obtained at different ages are summarised in Figs. B.21, B.22, B.23 and B.24 at 7 days, 28 days, T1 and T2 respectively.

Table B.5 S-B0 batch characterization: compressive strength test results

Specimen	Compressive strength f_c (MPa)				
	7 days	28 days	T1	T2	T3
S-B0-405	30.0	–	–	–	–
S-B0-406	31.3	–	–	–	–
S-B0-407	32.4	–	–	–	–
S-B0-408	–	37.4	–	–	–
S-B0-409	–	37.7	–	–	–
S-B0-410	–	38.0	–	–	–
Mean	**31.23**	**37.70**	–	–	–
CV (%)	3.8	0.8	–	–	–

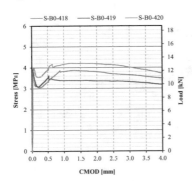

Specimen	f_L	$f_{R,1}$	$f_{R,2}$	$f_{R,3}$	$f_{R,4}$
S-B0-414	3.12	2.45	3.06	3.09	2.93
S-B0-415	3.49	3.16	3.77	3.62	3.61
S-B0-416	3.53	2.90	3.48	3.41	3.21
S-B0-417	3.35	2.07	2.37	2.46	2.50
Mean	**3.37**	**2.64**	**3.17**	**3.15**	**3.06**
CV [%]	5.5	18.2	19.2	16.1	15.3

Fig. B.21 S-B0 flexural residual strength characterization: test results at 7 days

Specimen	f_L	$f_{R,1}$	$f_{R,2}$	$f_{R,3}$	$f_{R,4}$
S-B0-418	3.85	3.31	3.85	3.70	3.56
S-B0-419	3.93	3.48	3.38	3.34	3.26
S-B0-420	3.99	3.89	4.21	4.13	3.87
Mean	**3.92**	**3.56**	**3.81**	**3.72**	**3.57**
CV [%]	1.9	8.3	10.9	10.7	8.5

Fig. B.22 S-B0 flexural residual strength characterization: test results at 28 days

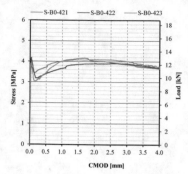

Specimen	f_L	$f_{R,1}$	$f_{R,2}$	$f_{R,3}$	$f_{R,4}$
S-B0-421	3.87	3.71	4.14	3.95	3.72
S-B0-422	4.18	3.40	3.86	3.90	3.79
S-B0-423	3.91	3.44	4.07	4.09	3.86
Mean	**3.99**	**3.52**	**4.02**	**3.98**	**3.79**
CV [%]	4.2	4.9	3.6	2.4	1.9

Fig. B.23 S-B0 flexural residual strength characterization: test results at T1

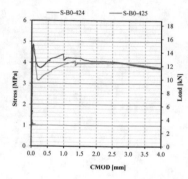

Specimen	f_L	$f_{R,1}$	$f_{R,2}$	$f_{R,3}$	$f_{R,4}$
S-B0-424	4.70	3.50	3.95	3.96	3.86
S-B0-425	4.86	4.06	4.21	4.05	3.80
Mean	**4.78**	**3.78**	**4.08**	**4.01**	**3.83**
CV [%]	2.4	10.4	4.4	1.5	1.1

Fig. B.24 S-B0 flexural residual strength characterization: test results at T2

Appendix C
Database and Parameter Definition

Aitor Llano-Torre and Pedro Serna

A convention regarding the data collection during the Round-Robin Test (RRT) was established to simplify the analysis of results. A first database proposal was discussed in the RILEM 261-CCF TC following the proposal found in previous publications [48, 49]. The final RRT database [18] was published under a Creative Commons license as supplementary material of this Round-Robin Test State-of-the-Art Report and it is available for the scientific community to improve the global knowledge in the long-term deformations of the cracked FRC specimens.

The relevant parameters were organized in ten groups divided in three main sections depending on the nature of the information (i.e. obtained directly from test, data calculated from test results or complementary data for further analysis). This classification made easier to compile the database containing all the specimens' information.

Although displacement notations are defined in terms of crack mouth opening displacement (CMOD), analogous notations also apply to the crack opening displacement (COD) or deflection (δ). The analogous symbols and definitions are obtained substituting CMOD by COD or δ.

A. Llano-Torre · P. Serna
Institute of Concrete Science and Technology ICITECH, Universitat Politècnica de València (UPV), Valencia, Spain
e-mail: aillator@upv.es

P. Serna
e-mail: pserna@cst.upv.es

© RILEM 2021

241

A. Llano-Torre and P. Serna (eds.), *Round-Robin Test on Creep Behaviour in Cracked Sections of FRC: Experimental Program, Results and Database Analysis*, RILEM State-of-the-Art Reports 34, https://doi.org/10.1007/978-3-030-72736-9

Table C.1 Groups of main parameters of the RRT database

Group	Description
A	Identity data
B	Environmental conditions during creep test
C	Concrete matrix characteristics (mean value of the series)
D	Fibres
E	Specimen dimensions
F	Pre-cracking test output data—Stage 1
G	Load level in Creep test
H	Creep test output data—Stage 2
I	Unloading stage of creep test output data

Global Data Obtained for Each Specimen Until Failure: Section 1

The first section is called "*Global Data Obtained for each Specimen until Failure*" and contains the parameters obtained in the test. Table C.1 shows the different groups of parameters presented in detail in Table C.2. Note that this information was requested for each specimen tested subjected to creep in the RRT.

Data for the Analysis Calculated Form Creep Test Data: Section 2

Table C.3 compiles in three groups the main creep parameters derived from the creep test results. Three coefficients were defined and calculated to compare the behaviour found in the RRT between laboratories: *creep coefficient referred to the creep stage* ($\varphi_{w,c}$), *creep coefficient referred to the origin of deformation* ($\varphi_{w,o}$) and *crack opening rate* (COR).

The creep coefficient referred to creep stage ($\varphi_{w,c}$) was calculated by means of the following Eq. (D.1):

$$\varphi_{w,c}^{j} = CMOD_{cd}^{j}/CMOD_{ci} \text{ where } CMOD_{cd}^{j} = CMOD_{ct}^{j} - CMOD_{ci} \quad (D.1)$$

The creep coefficient referred to creep origin ($\varphi_{w,o}$) was calculated by means of the following Eq. (D.2):

$$\varphi_{w,o}^{j} = CMOD_{cd}^{j}/CMOD_{oi} \text{ where } CMOD_{oi} = CMOD_{pr} + CMOD_{ci} \quad (D.2)$$

The crack opening rate (COR) was calculated by means of the following Eq. (D.3):

Table C.2 Global data for each specimen tested following the RRT procedure

Group	N°	Data	Description	Units
A	*Identity data*			
	A.1	Series	Batch of the specimen	–
	A.2	Name/Specimen	Name/Specimen	–
	A.3	Reference	Reference number (Series + Name/Number)	–
	A.4	Laboratory	Reference number of the Laboratory	–
B	*Environmental conditions during creep test*			
	B.1	T	Average temperature	°C
	B.2	T_{min}	Minimum temperature during creep test	°C
	B.3	T_{max}	Maximum temperature during creep test	°C
	B.4	HR	Average relative humidity	%
	B.5	HR_{min}	Minimum relative humidity during creep test	%
	B.6	HR_{max}	Maximum relative humidity during creep test	%
C	*Concrete matrix characteristics (mean value of the series)*			
	C.1	D_{max}	Aggregate maximum size diameter	mm
	C.2	f_{ck}/f_{ck}	Targeted strength/Concrete Class	MPa
	C.3	$f_{c,7}$	Compressive strength (Ø150 × 300 mm) at 7 days	MPa
	C.4	$f_{c,28}$	Compressive strength (Ø150 × 300 mm) at 28 days	MPa
	C.5	$f_{c,T1}$	Compressive strength (Ø150 × 300 mm) at start of creep test (T1)	MPa

(continued)

Table C.2 (continued)

Group	N°	Data	Description	Units
	C.6	$f_{c,T2}$	Compressive strength (Ø150 × 300 mm) at half creep test (T2)	MPa
	C.7	$f_{c,T3}$	Compressive strength (Ø150 × 300 mm) at the end of creep test (T3)	MPa
D	*Fibres*			
	D.1	Material	Material	–
	D.2	Brand	Brand/Commercial reference	–
	D.3	Length	Length	mm
	D.4	Diameter	Diameter	mm
	D.5	Slenderness	Slenderness	–
	D.6	E	Young modulus	GPa
	D.7	σ_y	Yield strength	MPa
	D.8	σ_u	Ultimate Tensile strength	MPa
	D.9	Dosage (weight)	Dosage in weight	kg/m^3
	D.10	Dosage (volume)	Dosage in volume	% Vol
E	*Specimen dimensions*			
	E.1	Shape	Specimen shape: prismatic, cylindrical, square or round panel	-
	E.2	Dimensions	Prismatic real dimensions by calliper (BxHxL) or cylindrical (ØDxH)	mm
	E.3	h_{notch}	Notch height	mm
	E.4	b_{notch}	Notch width	mm

(continued)

Table C.2 (continued)

Group	N°	Data	Description	Units
F	**Pre-cracking test output data—Stage 1**			
	F.1	3PBT/4PBT	**Flexural Beams**—Specify the adopted load configuration	-
	F.2	L	**Flexural Beams**—Support span	mm
	F.3	l_a	**Flexural Beams**—Distance between support and nearest loading point	mm
	F.4	l_b	**Flexural Beams**—Distance between loading points (only for 4PBT)	mm
	F.5	f_L	Residual strength at LOP	MPa
	F.6	$f_{R,p}$	Residual strength at $CMOD_p$	MPa
	F.7	F_L	Load at LOP	kN
	F.8	$F_{R,p}$	Load at $CMOD_p$	kN
	F.9	$CMOD/COD\delta$	Reference displacement measured and registered during tests	–
	F.10	$CMOD_{pn}$	Nominal pre-cracking level expected	μm
	F.11	$CMOD_p$	Real crack opening before unloading the specimen [**Point B**]	μm
	F.12	$CMOD_{pri}$	Elastic CMOD recovery when unloading the specimen [**Point C**]	μm
	F.13	$CMOD_{pr}$	Remain CMOD 10′ after unloading the specimen [**Point D**]	μm

(continued)

Table C.2 (continued)

Group	N°	Data	Description	Units
		3PBT **4PBT**		
G		*Load level in Creep test*		
	G.1	3PBT/4PBT	**Flexural Beams**—Specify the adopted load configuration	-
	G.2	L	**Flexural Beams**—Support span	mm
	G.3	l_a	**Flexural Beams**—Distance between support and nearest loading point	mm
	G.4	l_b	**Flexural Beams**—Distance between loading points (if 4PBT is in G.1)	mm
	G.5	I_n	Nominal Creep index [% of $f_{R,p}$]	%
	G.6	$F_{R,c}$	Load applied in creep test	kN
	G.7	$f_{R,c}$	Stress applied in creep test	MPa
	G.8	I_c	Creep index [$I_c = f_{R,c}/f_{R,p}$]	%
	G.9	t_{ci}	Time in which load has been applied (from start loading to $f_{R,c}$)	seconds
	G.10	t_f, t_{max}	Total time that load was sustained in the creep test	days

(continued)

Table C.2 (continued)

Group	N°	Data	Description	Units
		3PBT **4PBT**		
H		*Creep test output data—Stage 2 (see point "N. Values of delayed displacement" in Table C.6)*		
	H.1	$CMOD_{ci}$	Instantaneous CMOD immediately after reaching $f_{R,c}$[**Point E**]	μm
	H.2	$CMOD_{ci}^{10'}$	Short-term CMOD 10′ after reaching $f_{R,c}$[**Point E + 10′**]	μm
	H.3	$CMOD_{ci}^{30'}$	Short-term CMOD 30′ after reaching $f_{R,c}$[**Point E + 30′**]	μm
	H.4	$CMOD_{ct}^{1}$	Total CMOD from point "D" at 1 day	μm
	H.5	$CMOD_{ct}^{2}$	Total CMOD from point "D" at 2 day	μm
	H.6	$CMOD_{ct}^{3}$	Total CMOD from point "D" at 3 day	μm
	H.7	$CMOD_{ct}^{5}$	Total CMOD from point "D" at 5 day	μm
	H.8	$CMOD_{ct}^{7}$	Total CMOD from point "D" at 7 day	μm
	H.9	$CMOD_{ct}^{14}$	Total CMOD from point "D" at 14 day	μm
	H.10	$CMOD_{ct}^{30}$	Total CMOD from point "D" at 30 day	μm
	H.11	$CMOD_{ct}^{60}$	Total CMOD from point "D" at 60 day	μm

(continued)

Table C.2 (continued)

Group	N°	Data	Description	Units
	H.12	$CMOD_{ct}^{90}$	Total CMOD from point "D" at 90 day	μm
	H.13	$CMOD_{ct}^{120}$	Total CMOD from point "D" at 120 day	μm
	H.14	$CMOD_{ct}^{150}$	Total CMOD from point "D" at 150 day	μm
	H.15	$CMOD_{ct}^{180}$	Total CMOD from point "D" at 180 day	μm
	H.16	$CMOD_{ct}^{210}$	Total CMOD from point "D" at 210 day	μm
	H.17	$CMOD_{ct}^{240}$	Total CMOD from point "D" at 240 day	μm
	H.18	$CMOD_{ct}^{270}$	Total CMOD from point "D" at 270 day	μm
	H.19	$CMOD_{ct}^{300}$	Total CMOD from point "D" at 300 day	μm
	H.20	$CMOD_{ct}^{330}$	Total CMOD from point "D" at 330 day	μm
	H.21	$CMOD_{ct}^{360}$	Total CMOD from point "D" at 360 day	μm

(continued)

Table C.2 (continued)

Group	N°	Data	Description	Units
I	*Unloading stage of creep test output data*			
	I.1	t_{cri}	Time to remove the load of the creep test	seconds
	I.2	$CMOD_{cri}$	Total CMOD immediately after unloading (load = 0 kN) **[Point H]**	μm
	I.3	$CMOD_{crd}$	Total CMOD 30 days after unloading **[Point G]**	μm
	I.4	t_{crd}	Time after unloading in which recovery was registered	days

Group	N°	Data	Description	Units
J	*Residual behaviour after creep (previously named "Flexural behaviour")*			
	J.1	$f_{PostCreep,R,2}$	Residual strength after creep at CMOD2	MPa
	J.2	$f_{PostCreep,R,3}$	Residual strength after creep at CMOD3	MPa
	J.3	$f_{PostCreep,R,4}$	Residual strength after creep at CMOD4	MPa

(continued)

Table C.2 (continued)

Group	N°	Data	Description	Units
	J.4	$f_{PostCreep,R,3}/f_{R,1}$	Post-crack behaviour	
	J.5	$F_{PostCreep,R,2}$	Load at CMOD2 after creep	kN
	J.6	$F_{PostCreep,R,3}$	Load at CMOD3 after creep	kN
	J.7	$F_{PostCreep,R,4}$	Load at CMOD4 after creep	kN

Table C.3 Groups of main parameters calculated for each specimen from creep test

Group	Description
K	Creep coefficients referred to creep stage
L	Creep coefficients referred to origin (with residual crack opening)
M	Crack opening rate

$$\mathrm{COR}^{\,j-k} = \left(\mathrm{CMOD}_{cd}^{k} - \mathrm{CMOD}_{cd}^{j}\right)/((k-j)/365) \qquad (D.3)$$

Table C.4 presents all the calculated creep coefficients and CORs for different time lapses. These values are available for each specimen in the RRT database.

Table C.4 Parameters calculated for each specimen for the creep test

Group	N°	Data	Description	Units
K	**Creep coefficients referred to creep stage**			
	K.1	$\varphi_{w,c}{}^{7}$	Creep coefficient referred to creep at 7 days	–
	K.2	$\varphi_{w,c}{}^{14}$	Creep coefficient referred to creep at 14 days	–
	K.3	$\varphi_{w,c}{}^{30}$	Creep coefficient referred to creep at 30 days	–
	K.4	$\varphi_{w,c}{}^{60}$	Creep coefficient referred to creep at 60 days	–
	K.5	$\varphi_{w,c}{}^{90}$	Creep coefficient referred to creep at 90 days	–
	K.6	$\varphi_{w,c}{}^{120}$	Creep coefficient referred to creep at 120 days	–
	K.7	$\varphi_{w,c}{}^{150}$	Creep coefficient referred to creep at 150 days	–
	K.8	$\varphi_{w,c}{}^{180}$	Creep coefficient referred to creep at 180 days	–
	K.9	$\varphi_{w,c}{}^{210}$	Creep coefficient referred to creep at 210 days	–
	K.10	$\varphi_{w,c}{}^{240}$	Creep coefficient referred to creep at 240 days	–
	K.11	$\varphi_{w,c}{}^{270}$	Creep coefficient referred to creep at 270 days	–
	K.12	$\varphi_{w,c}{}^{300}$	Creep coefficient referred to creep at 300 days	–
	K.13	$\varphi_{w,c}{}^{330}$	Creep coefficient referred to creep at 330 days	–
	K.14	$\varphi_{w,c}{}^{360}$	Creep coefficient referred to creep at 360 days	–
L	**Creep coefficients referred to origin**			
	L.1	$\varphi_{w,o}{}^{7}$	Creep coefficient referred to origin at 7 days	–
	L.2	$\varphi_{w,o}{}^{14}$	Creep coefficient referred to origin at 14 days	–
	L.3	$\varphi_{w,o}{}^{30}$	Creep coefficient referred to origin at 30 days	–
	L.4	$\varphi_{w,o}{}^{60}$	Creep coefficient referred to origin at 60 days	–
	L.5	$\varphi_{w,o}{}^{90}$	Creep coefficient referred to origin at 90 days	–
	L.6	$\varphi_{w,o}{}^{120}$	Creep coefficient referred to origin at 120 days	–
	L.7	$\varphi_{w,o}{}^{150}$	Creep coefficient referred to origin at 150 days	–
	L.8	$\varphi_{w,o}{}^{180}$	Creep coefficient referred to origin at 180 days	–
	L.9	$\varphi_{w,o}{}^{210}$	Creep coefficient referred to origin at 210 days	–

(continued)

Table C.4 (continued)

Group	N°	Data	Description	Units
	L.10	$\varphi_{w,o}{}^{240}$	Creep coefficient referred to origin at 240 days	–
	L.11	$\varphi_{w,o}{}^{270}$	Creep coefficient referred to origin at 270 days	–
	L.12	$\varphi_{w,o}{}^{300}$	Creep coefficient referred to origin at 300 days	–
	L.13	$\varphi_{w,o}{}^{330}$	Creep coefficient referred to origin at 330 days	–
	L.14	$\varphi_{w,o}{}^{360}$	Creep coefficient referred to origin at 360 days	–
M	**Crack opening rate (COR)**			
	M.1	COR^{0-30}	Crack opening rate from 0 to 30 days	μm/year
	M.2	COR^{30-60}	Crack opening rate from 30 to 60 days	μm/year
	M.3	COR^{60-90}	Crack opening rate from 60 to 90 days	μm/year
	M.4	COR^{90-120}	Crack opening rate from 90 to 120 days	μm/year
	M.5	$COR^{120-150}$	Crack opening rate from 120 to 150 days	μm/year
	M.6	$COR^{150-180}$	Crack opening rate from 150 to 180 days	μm/year
	M.7	$COR^{180-210}$	Crack opening rate from 180 to 210 days	μm/year
	M.8	$COR^{210-240}$	Crack opening rate from 210 to 240 days	μm/year
	M.9	$COR^{240-270}$	Crack opening rate from 240 to 270 days	μm/year
	M.10	$COR^{270-300}$	Crack opening rate from 270 to 300 days	μm/year
	M.11	$COR^{300-330}$	Crack opening rate from 300 to 330 days	μm/year
	M.12	$COR^{330-360}$	Crack opening rate from 330 to 360 days	μm/year

Table C.5 Main parameters calculated for each specimen from creep test

Group	Description
C	Concrete matrix characteristics (mean value of the serie)
E	Specimen dimensions
N	Values of delayed displacement origin
O	Fibre Counting

Complementary Data for Analysis: Section 3

The third section, called "Complementary Data for Analysis", compiles complementary data for the previously defined groups seen in Table C.5. The description and units of the specific data required for each group are defined in Table C.6.

Table C.6 Complementary creep test data requested for each specimen

Group	N°	Data	Description	Units
C	**Concrete matrix characteristics (mean value of the serie)**			
	C.9	Curing	Time of curing before creep test	days
	C.10	Curing conditions	Not controlled, moist chamber…	–
E	**Specimen dimensions**			
	E.5	Notch shape	Shape of the notch: rectangular, semicircular, pointed…	–
	E.6	Age at notch	Age of specimen when notching was done	days
	E.7		Time between notching and pre-cracking	days
	E.8		After notch, specimen kept at controlled conditions?	Yes/No
N	**Values of delayed displacement**			
	N.1	Delayed values	Indicate if provided data are real values or trend line values	Real/Trend
	N.2	Trend line	Define type of trend line used to fit creep curve	–
O	**Fibre Counting**			
	O.1	[FB] Side A—Top	Flexural beams—Fibres in the upper third of the side A	fibres/cm^2
	O.2	[FB] Side A—Middle	Flexural beams—Fibres in the central third of the side A	fibres/cm^2
	O.3	[FB] Side A—Bottom	Flexural beams—Fibres in the lower third of the side A	fibres/cm^2
	O.4	[FB] Side B—Top	Flexural beams—Fibres in the upper third of the side B	fibres/cm^2
	O.5	[FB] Side B—Middle	Flexural beams—Fibres in the central third of the side B	fibres/cm^2
	O.6	[FB] Side B—Bottom	Flexural beams—Fibres in the lower third of the side B	fibres/cm^2
	O.7	[DT] Side A	Direct Tension—Fibres in the side A	fibres/cm^2
	O.8	[DT] Side B	Direct Tension—Fibres in the side B	fibres/cm^2

Appendix D
Creep Coefficients Results

Aitor Llano-Torre, Pedro Serna and Clementina del Prete

The Appendix D shows the evolution of creep coefficients referred to both creep stage and origin of deformations obtained for all participant laboratories. Figures show the creep coefficients for both SyFRC and SFRCs in a box and whisker plot, as well as the median evolution over time. Additional predictions of the creep coefficient by using the model in Eq. (8.10) are also presented.

Creep Coefficients Referred to Creep Stage

Flexural Tests

As complementary data, this section compiles the evolution in time of the creep coefficients referred to the creep stage ($\varphi_{w,c}$) obtained by means of the Eq. (D.1) for the different participant laboratories. In those cases where the three different $CMOD_{ci}$ references described in this report ($CMOD_{ci}$, $CMOD_{ci}{}^{10'}$ and $CMOD_{ci}{}^{30'}$) were reported, the creep coefficient evolution is depicted for each $CMOD_{ci}$ reference

A. Llano-Torre · P. Serna
Institute of Concrete Science and Technology ICITECH, Universitat Politècnica de València (UPV), Valencia, Spain
e-mail: aillator@upv.es

P. Serna
e-mail: pserna@cst.upv.es

C. del Prete
Department of Civil, Chemical, Environmental and Materials Engineering DICAM, University of Bologna, Bologna, Italy
e-mail: clementina.delprete2@unibo.it

© RILEM 2021
A. Llano-Torre and P. Serna (eds.), *Round-Robin Test on Creep Behaviour in Cracked Sections of FRC: Experimental Program, Results and Database Analysis*, RILEM State-of-the-Art Reports 34, https://doi.org/10.1007/978-3-030-72736-9

255

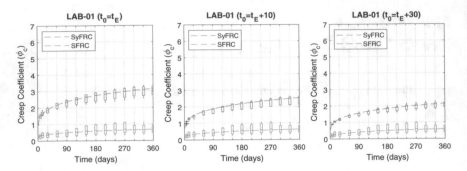

Fig. D.1 Creep coefficients referred to creep stage ($\varphi_{w,c}$) calculated from LAB-01 results

Fig. D.2 Creep coefficients referred to creep stage ($\varphi_{w,c}$) calculated from LAB-02 results

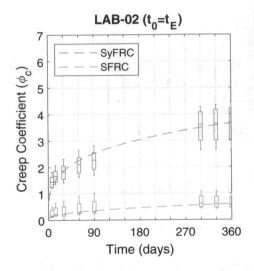

for comparative purposes (Figs. D.1, D.2, D.3, D.4, D.5, D.6, D.7, D.8, D.9, D.10, D.11 and D.12).

Direct Tension Tests

See Figs. D.13 and D.14.

Square Panel Tests

See Figs. D.15 and D.16.

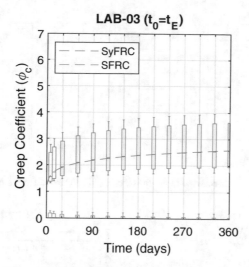

Fig. D.3 Creep coefficients referred to creep stage ($\varphi_{w,c}$) calculated from LAB-03 results

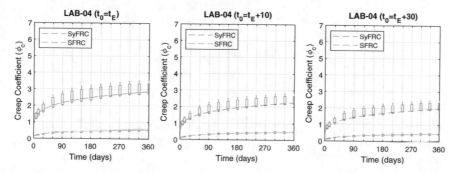

Fig. D.4 Creep coefficients referred to creep stage ($\varphi_{w,c}$) calculated from LAB-04 results

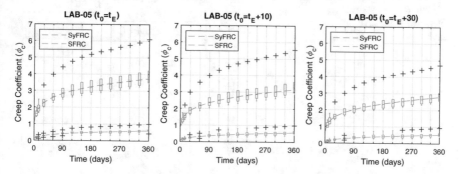

Fig. D.5 Creep coefficients referred to creep stage ($\varphi_{w,c}$) calculated from LAB-05 results

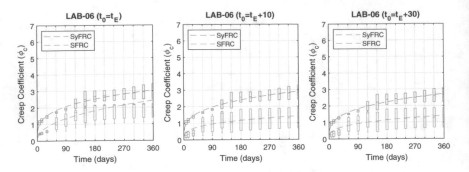

Fig. D.6 Creep coefficients referred to creep stage ($\varphi_{w,c}$) calculated from LAB-06 results

Fig. D.7 Creep coefficients referred to creep stage ($\varphi_{w,c}$) calculated from LAB-07 results

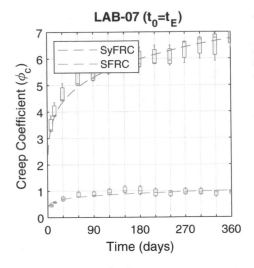

Fig. D.8 Creep coefficients referred to creep stage ($\varphi_{w,c}$) calculated from LAB-08 results

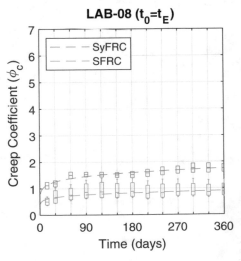

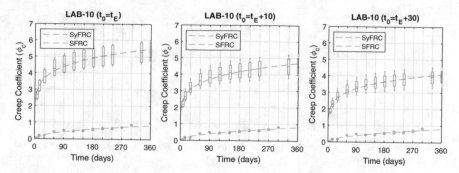

Fig. D.9 Creep coefficients referred to creep stage ($\varphi_{w,c}$) calculated from LAB-10 results

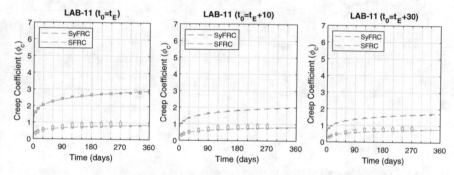

Fig. D.10 Creep coefficients referred to creep stage ($\varphi_{w,c}$) calculated from LAB-11 results

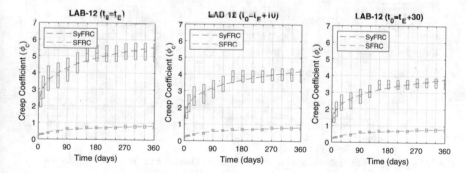

Fig. D.11 Creep coefficients referred to creep stage ($\varphi_{w,c}$) calculated from LAB-12 results

Round Panel Tests

See Fig. D.17.

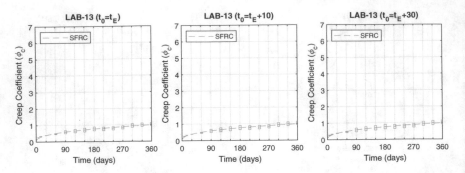

Fig. D.12 Creep coefficients referred to creep stage ($\varphi_{w,c}$) calculated from LAB-13 results

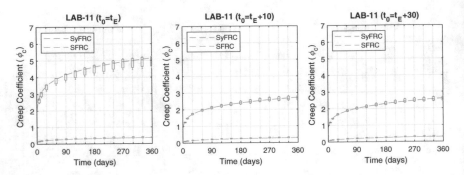

Fig. D.13 Creep coefficients referred to creep stage ($\varphi_{w,c}$) calculated from LAB-11 results

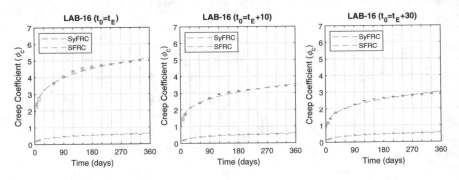

Fig. D.14 Creep coefficients referred to creep stage ($\varphi_{w,c}$) calculated from LAB-16 results

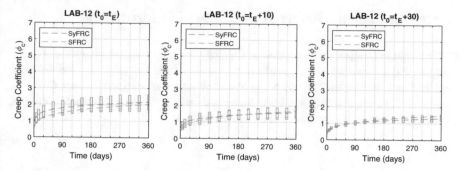

Fig. D.15 Creep coefficients referred to creep stage ($\varphi_{w,c}$) calculated from LAB-12 results

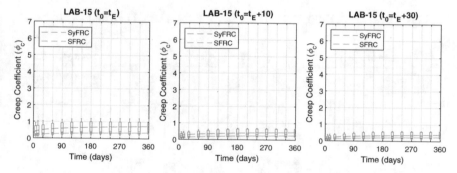

Fig. D.16 Creep coefficients referred to creep stage ($\varphi_{w,c}$) calculated from LAB-15 results

Fig. D.17 Creep coefficients referred to creep stage ($\varphi_{w,c}$) calculated from LAB-17 results

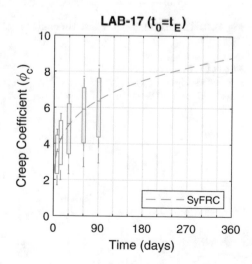

Creep Coefficients Referred to Origin of Deformations

This section compiles the evolution in time of the creep coefficients referred to the origin of deformations ($\varphi_{w,o}$) obtained by means of the Eq. (D.2) for the different participant laboratories. The creep coefficient evolution is depicted for each $CMOD_{oi}$ reference for comparative purposes in those cases where the three different short-term $CMOD_{ci}$ references were reported by the participant.

Flexural Tests

See Figs. D.18, D.19, D.20, D.21, D.22, D.23, D.24, D.25, D.26 and D.27.

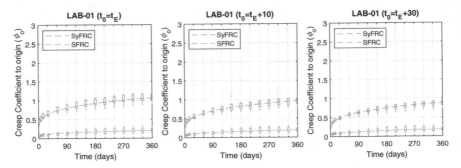

Fig. D.18   Creep coefficients referred to origin ($\varphi_{w,o}$) calculated from LAB-01 results

Fig. D.19   Creep coefficients referred to origin ($\varphi_{w,o}$) calculated from LAB-02 results

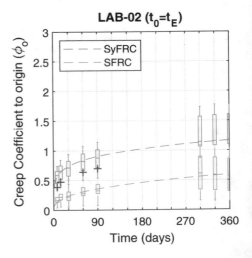

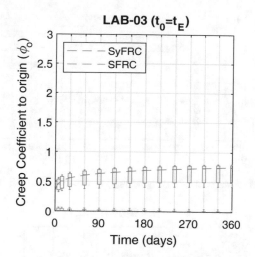

Fig. D.20 Creep coefficients referred to origin ($\varphi_{w,o}$) calculated from LAB-03 results

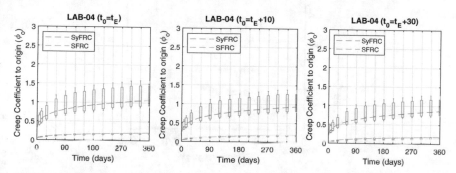

Fig. D.21 Creep coefficients referred to origin ($\varphi_{w,o}$) calculated from LAB-04 results

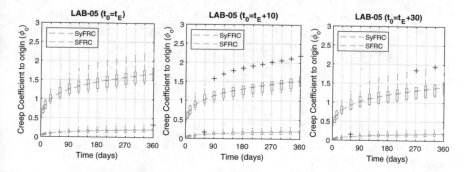

Fig. D.22 Creep coefficients referred to origin ($\varphi_{w,o}$) calculated from LAB-05 results

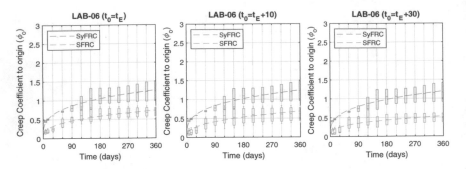

Fig. D.23 Creep coefficients referred to origin ($\varphi_{w,o}$) calculated from LAB-06 results

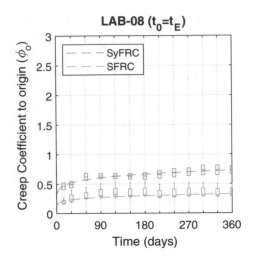

Fig. D.24 Creep coefficients referred to origin ($\varphi_{w,o}$) calculated from LAB-08 results

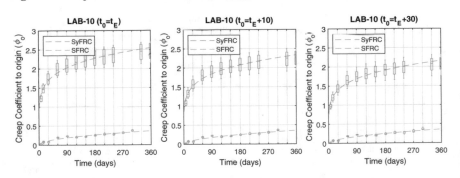

Fig. D.25 Creep coefficients referred to origin ($\varphi_{w,o}$) calculated from LAB-10 results

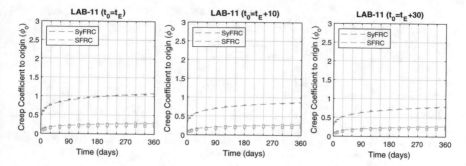

Fig. D.26 Creep coefficients referred to origin ($\varphi_{w,o}$) calculated from LAB-11 results

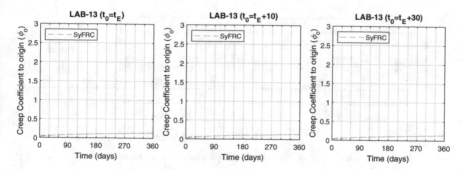

Fig. D.27 Creep coefficients referred to origin ($\varphi_{w,o}$) calculated from LAB-13 results

Direct Tension Tests

See Figs. D.28 and D.29.

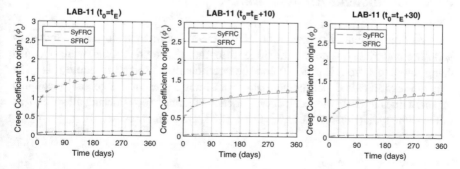

Fig. D.28 Creep coefficients referred to origin ($\varphi_{w,o}$) calculated from LAB-11 results

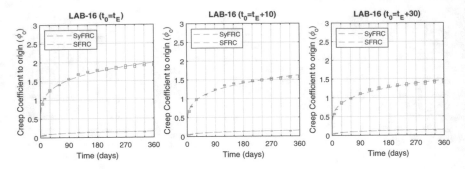

Fig. D.29 Creep coefficients referred to origin ($\varphi_{w,o}$) calculated from LAB-16 results

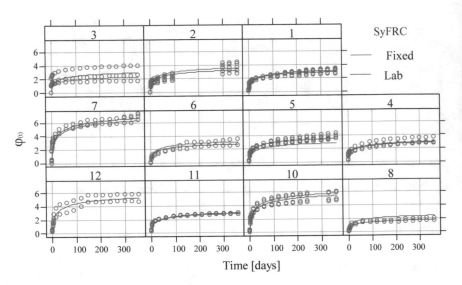

Fig. D.30 Prediction of the creep coefficient for each laboratory, considering $t_0 = t_E$ and SyFRC specimens

Creep Coefficients Analysis Annex

This section presents the prediction of the creep coefficients for each laboratory and fibre material considering the three different t_0. The predictions were obtained using the model from Eq. (8.10) and the coefficients presented in Table 8.11. The blue circles indicate experimental data, cyan lines represent the general mean and magenta lines indicate the mean curve for each laboratory considering all specimens of the same type (Figs. D.30, D.31, D.32, D.33, D.34 and D.35).

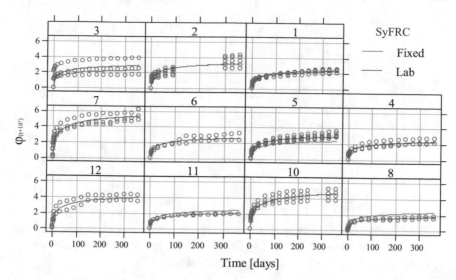

Fig. D.31 Prediction of the creep coefficient for each laboratory, considering $t_0 = t_E + 10'$ and SyFRC specimens

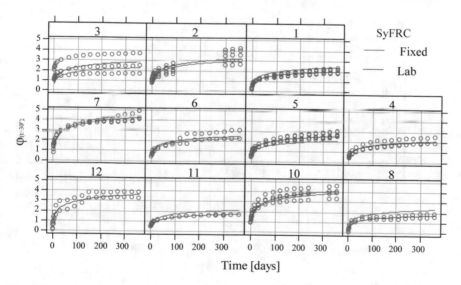

Fig. D.32 Prediction of the creep coefficient for each laboratory, considering $t_0 = t_E + 30'$ and SyFRC specimens

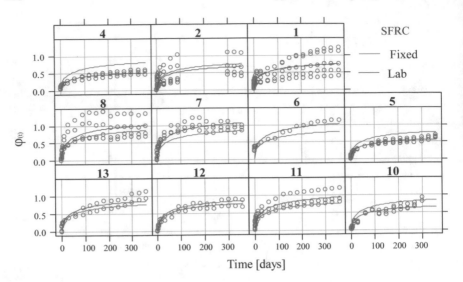

Fig. D.33 Prediction of the creep coefficient for each laboratory, considering $t_0 = t_E$ and SFRC specimens

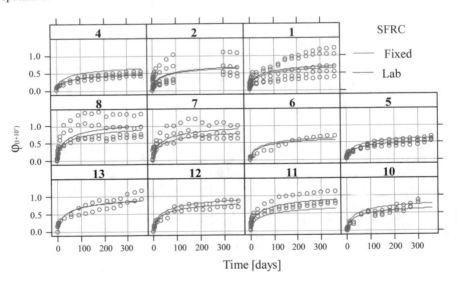

Fig. D.34 Prediction of the creep coefficient for each laboratory, considering $t_0 = t_E + 10'$ and SFRC specimens

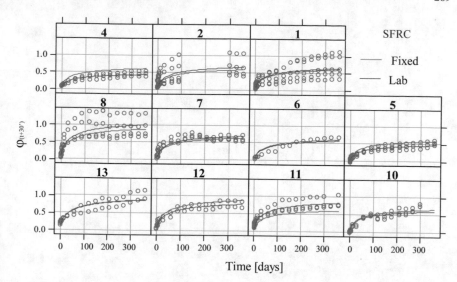

Fig. D.35 Prediction of the creep coefficient for each laboratory, considering $t_0 = t_E + 30'$ and SFRC specimens

Appendix E
Crack Opening Rate (COR) Results

Aitor Llano-Torre and Pedro Serna

The Appendix E collects the crack opening rate (COR) evolution and mean curves derived from the results of each participant laboratory. Figures show the COR obtained in 30-day time lapses for both macro-synthetic fibre-reinforced concrete (SyFRC) and steel fibre-reinforced concrete (SFRC) specimens in a box and whisker plot and the time evolution of median COR. The COR values were obtained by means of the Eq. (8.11) as defined in Sect. 8.5.

Flexural Tests

See Figs. E.1, E.2, E.3, E.4, E.5, E.6, E.7, E.8, E.9, E.10, E.11 and E.12.

Direct Tension Tests

See Figs. 13 and 14.

A. Llano-Torre · P. Serna
Institute of Concrete Science and Technology ICITECH, Universitat Politècnica de València (UPV), Valencia, Spain
e-mail: aillator@upv.es

P. Serna
e-mail: pserna@cst.upv.es

© RILEM 2021
A. Llano-Torre and P. Serna (eds.), *Round-Robin Test on Creep Behaviour in Cracked Sections of FRC: Experimental Program, Results and Database Analysis*, RILEM State-of-the-Art Reports 34, https://doi.org/10.1007/978-3-030-72736-9

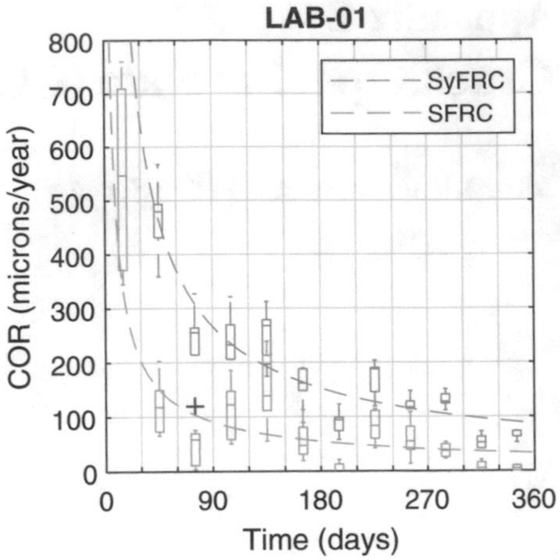

Fig. E.1 COR calculated
from creep test results by
LAB-01

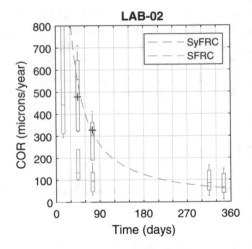

Fig. E.2 COR calculated
from creep test results by
LAB-02

Square Panel Tests

See Figs. E.15, E.16 and E.17.

Fig. E.3 COR calculated from creep test results by LAB-03

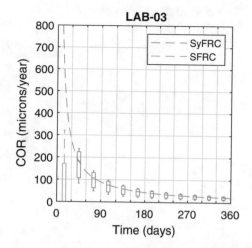

Fig. E.4 COR calculated from creep test results by LAB-04

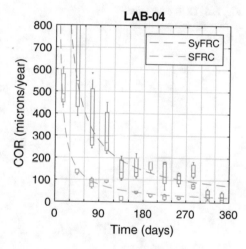

Fig. E.5 COR calculated
from creep test results by
LAB-05

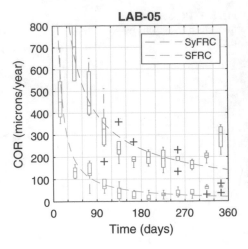

Fig. E.6 COR calculated
from creep test results by
LAB-06

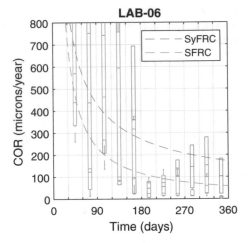

Fig. E.7 COR calculated from creep test results by LAB-07

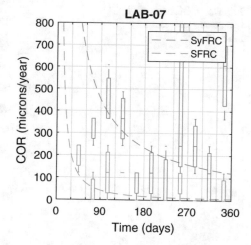

Fig. E.8 COR calculated from creep test results by LAB-08

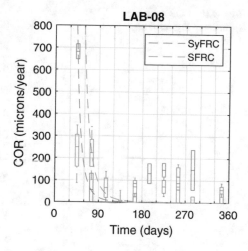

Fig. E.9 COR calculated
from creep test results by
LAB-10

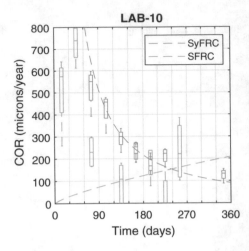

Fig. E.10 COR calculated
from creep test results by
LAB-11

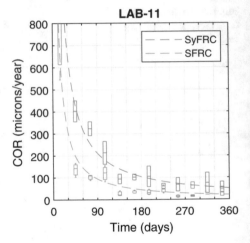

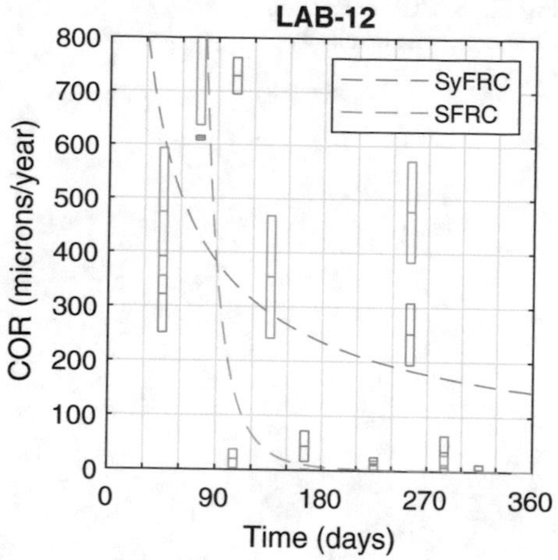

Fig. E.11 COR calculated from creep test results by LAB-12

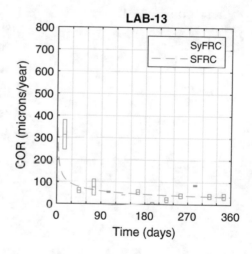

Fig. E.12 COR calculated from creep test results by LAB-13

Fig. E.13 COR calculated
from creep test results by
LAB-11

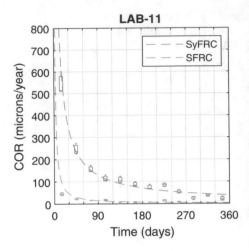

Fig. E.14 COR calculated
from creep test results by
LAB-16

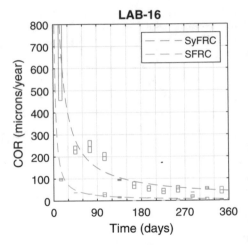

Fig. E.15 COR calculated
from creep test results by
LAB-12

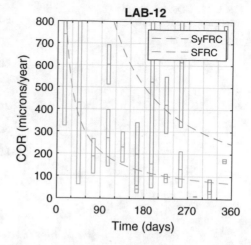

Fig. E.16 COR calculated
from creep test results by
LAB-15

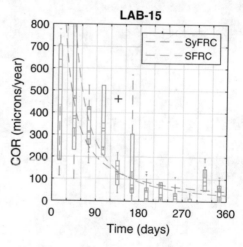

Fig. E.17 COR calculated
from creep test results by
LAB-18

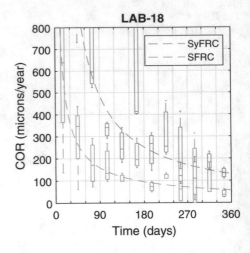

References

1. *fib* Model Code for Concrete Structures 2010 (2013), Wiley-VCH Verlag GmbH & Co. KGaA, pp. 74–150.
2. ACI Committee 318, American Concrete Institute, (2014). Building code requirements for structural concrete (ACI 318–14): An ACI standard.
3. Monetti, D. H., Llano-Torre, A., Torrijos, M. C., Giaccio, G., Zerbino, R., Martí-Vargas, J. R., & Serna, P. (2019). Long-term behavior of cracked fiber reinforced concrete under service conditions. Construction and Building Materials, 196, 649–658. https://doi.org/10.1016/j.con buildmat.2018.10.230.
4. Serna P., Llano-Torre A. and Cavalaro S. H. P. (ed) 2017 Creep behaviour in cracked sections of fibre reinforced concrete: Proceedings of the International RILEM Workshop FRC-CREEP 2016 (RILEM bookseries 14 (Dordrecht: Springer).
5. Zerbino, R., Monetti, D.H. & Giaccio, G. Creep behaviour of cracked steel and macro-synthetic fibre reinforced concrete. Mater Struct 49, 3397–3410 (2016). https://doi.org/10.1617/s11527-015-0727-y.
6. Arango, S., Serna, P., Martí-Vargas, J. R. & García-Taengua, E. (2012), A Test Method to Characterize Flexural Creep Behaviour of Pre-cracked FRC Specimens. Experimental Mechanics, 52, pp. 1067–1078.
7. Zerbino R. & Barragán B. (2012), Long-term behavior of cracked steel fiber reinforced concrete beams under sustained loading. ACI Materials Journal, 109 (2), pp. 215–224.
8. Kusterle W. - Viscous material behavior of solids- creep of polymer fiber reinforced concrete, 5th Central European Congress on Concrete Engineering, Baden (2009).
9. García-Taengua, E., Arango, S., Martí-Vargas, J. R. & Serna, P. (2014), Flexural creep of steel fiber reinforced concrete in the cracked state. Construction and Building Materials, 65, pp. 321–329.
10. Larive C., Rogat D., Chamoley D., Welby N., Regnard A., 2015. "Creep behaviour of fibre reinforced sprayed concrete", SEE Tunnel:Promoting Tunneling in SEE Region, ITA WTC 2015 Congress and 41st General Assembly May 22-28, 2015, Dubrovnik, Croatia.
11. Bernard E.S., "Influence of fiber type on creep deformation of cracked fiber-reinforced shotcrete panels, ACI MATERIALS JOURNAL, Title no. 107-M54, 474–480, (2010).
12. Babafemi, A. J., & Boshoff, W. P. (2015). Tensile creep of macro-synthetic fibre reinforced concrete (MSFRC) under uni-axial tensile loading. Cement and Concrete Composites, 55, 62–69. https://doi.org/10.1016/j.cemconcomp.2014.08.002.
13. Vrijdaghs, R., di Prisco, M. & Vandewalle, L. Uniaxial tensile creep of a cracked polypropylene fiber reinforced concrete. *Mater Struct* **51**, 5 (2018). https://doi.org/10.1617/s11527-017-1132-5.

© RILEM 2021
A. Llano-Torre and P. Serna (eds.), *Round-Robin Test on Creep Behaviour in Cracked Sections of FRC: Experimental Program, Results and Database Analysis*, RILEM State-of-the-Art Reports 34, https://doi.org/10.1007/978-3-030-72736-9

14. di Prisco, M.; Ferrara, L. & Lamperti, M. G. L. (2013), Double edge wedge splitting (DEWS): an indirect tension test to identify post-cracking behaviour of fibre reinforced cementitious composites Materials and Structures, 2013, 46, 1893–1918.

15. Buratti N. & Mazzotti C., (2012). Effects of different types and dosages of fibres on the long-term behaviour of fibre-reinforced selfcompacting concrete. In: Proc. 8th RILEM Int. Symp. on Fibre Reinforced Concrete: challenges and opportunities [BEFIB 2012] (pp. 715–725). Guimaraes, Portugal.

16. Vasanelli, E., Micelli, F., Aiello, M. A. & Plizzari, G. (2013), Long term behavior of FRC flexural beams under sustained load. Engineering Structures, 56, pp. 1858–1867.

17. Llano-Torre A., Serna P., Cavalaro S.H.P., 'International Round Robin Test on creep behavior of FRC supported by the RILEM TC 261-CCF'. In: Proceedings of the BEFIB 2016, 9th RILEM International Symposium on Fiber Reinforced Concrete, pp. 127–140, Vancouver, Canada, 19–21 Sept 2016.

18. Llano-Torre, A., Serna Ros, P., Cavalaro, S., Kusterle, W., Moro, S., Zerbino, RL., Gettu, R., Pauwels H., Nishiwaki T., Parmentier B., Buratti N., Toledo Filho R.D., Charron J.P., Larive C., Boshoff W.P., Bernard, E.S. and Kompatscher M. (2021). Database of the Round-Robin Test on Creep Behaviour in Cracked Sections of Fibre Reinforced Concrete organised by the RILEM Technical Committee 261-CCF. https://doi.org/10.4995/Dataset/10251/163221.

19. EN 14845–1:2007. Test methods for fibres in concrete - Part 1: Reference concretes, CEN - European Committee for Standardization, Brussels (2007).

20. EN 12390–3:2009, Testing hardened concrete - Part 3: Compressive strength of test specimens, CEN - European Committee for Standardization, Brussels (2009).

21. EN 12390–13:2013, Testing hardened concrete - Part 13: Determination of secant modulus of elasticity in compression, CEN - European Committee for Standardization, Brussels (2013).

22. EN 14651:2005+A1:2007. Test Method for Metallic Fibre Concrete - Measuring the Flexural tensile Strength (Limit of Proportionality (LOP), Residual), CEN - European Committee for Standardization, Brussels (2005).

23. ASTM C512 / C512M-15, Standard Test Method for Creep of Concrete in Compression, ASTM International, West Conshohocken, PA, 2015, DOI: https://doi.org/10.1520/C0512_C0512M-15, www.astm.org.

24. OBV (Austrian Society for Construction Technology): Guideline Fibre Reinforced Concrete (in German: Richtlinie Faserbeton), Vienna (2008).

25. DAfStb: Guideline steel fibre reinfoced concrete (in German Richtlinie Stahlfaserbeton). Beuth-Verlag, Berlin (1996).

26. EFNARC, Testing Sprayed Concrete – Creep Test on Square Panel, June 2012.

27. Fascicule Asquapro « Utilisation des fibres pour le renforcement des bétons projetés de soutènement provisoire des tunnels », 2014, 51 p.

28. EN 14488–5:2006. Testing sprayed concrete - Part 5: Determination of energy absorption capacity of fibre reinforced slab specimens, Residual), CEN - European Committee for Standardization, Brussels (2006).

29. ASTM C1550-12a, Standard Test Method for Flexural Toughness of Fiber Reinforced Concrete (Using Centrally Loaded Round Panel), ASTM International, West Conshohocken, PA, 2012, DOI: https://doi.org/10.1520/C1550-12A, www.astm.org.

30. Llano-Torre A., Arango S.E., García-Taengua E., Martí-Vargas J.R., Serna P. (2017) Influence of Fibre Reinforcement on the Long-Term Behaviour of Cracked Concrete. In: Serna P., Llano-Torre A., Cavalaro S. (eds) Creep Behaviour in Cracked Sections of Fibre Reinforced Concrete. RILEM Bookseries, vol 14. Springer, Dordrecht. https://doi.org/10.1007/978-94-024-1001-3_16.

31. 31. A. Llano-Torre, P. Serna, R. Zerbino, J.R. Martí-Vargas, Effect of initial crack opening on flexural creep behavior of FRC specimens, RILEM Proceedings PRO 116 - 9th RILEM International Symposium on Fiber Reinforced Concrete: The Modern Landscape (BEFIB 2016), Vancouver, 2016, pp. 117–126.

32. Bast, T.; Eder, A.; Kusterle, W.: Kriechversuche an Kunststoffmakrofaserbetonen. Untersuchungen zum Langzeitverhalten von Faserbetonen unter Biegezugbeanspruchung – ein

Zwischenbericht. 11. Vilser Baustofftag, Reutte, 15. 3. 2007, Zement + Beton Handels- und Werbeges. m. b. H., Vienna.

33. Kusterle, W.: Creep of Fibre Reinforced Concrete – Flexural Test on Beams. Proc. Fibre Concrete 2015, CTU in Prague, Faculty of Civil Engineering, September 10–11. 2015, Prague.

34. Kusterle W. (2017) Flexural Creep Tests on Beams—8 Years of Experience with Steel and Synthetic Fibres. In: Serna P., Llano-Torre A., Cavalaro S. (eds) Creep Behaviour in Cracked Sections of Fibre Reinforced Concrete. RILEM Bookseries, vol 14. Springer, Dordrecht.

35. Zerbino, R.L., Giaccio G.M., Monetti D.H. and Torrijos M.C. "Effect of beam width on the creep behaviour of cracked fibre reinforced concrete" en Creep Behaviour in Cracked Sections of Fibre Reinforced Concrete, Proc. International RILEM Workshop FRC-CREEP 2016, Eds: P. Serna, A. Llano-Torre, S.H.P. Cavalaro, RILEM Bookseries V14, Springer, Valencia, 2016, pp 169-178.

36. Nishiwaki T., Kwon S., Otaki H., Igarashi G., Shaikh F.U., Fantilli A.P. (2017) Experimental Study on Time-Dependent Behavior of Cracked UHP-FRCC Under Sustained Loads. In: Serna P., Llano-Torre A., Cavalaro S. (eds) Creep Behaviour in Cracked Sections of Fibre Reinforced Concrete. RILEM Bookseries, vol 14. Springer, Dordrecht.

37. Buratti N., Mazzotti C. (2016) Experimental tests on the long-term behaviour of SFRC and MSFRC in bending and direct tension. BEFIB 2016 - 9th RILEM International Symposium on Fiber Reinforced Concrete. Vancouver, Canada, 19–21 September.

38. Buratti N., Mazzotti C. (2017) Creep Testing Methodologies and Results Interpretation. In: Serna P., Llano-Torre A., Cavalaro S. (eds) Creep Behaviour in Cracked Sections of Fibre Reinforced Concrete. RILEM Bookseries, vol 14. Springer, Dordrecht. https://doi.org/10.1007/978-94-024-1001-3_2.

39. 39. Daviau-Desnoyers D, Charron J-P, Massicotte B, Rossi P, Tailhan J-L. (2016). Characterization of the Propagation of a Macrocrack under Sustained Loading in Steel Fibre Reinforced Concrete. Materials and Structures. Volume 49, Issue 3, 969–982. https://doi.org/10.1617/s11527-015-0552-3.

40. C. Larive, D. Rogat, D. Chamoley, A. Regnard, T. Pannetier, C. Thuaud. Influence of fibres on the creep behaviour of reinforced sprayed concrete. In: Proceedings of the World Tunnel Congress 2016 (WTC 2016), San Francisco, California, USA, 22–28 April 2016. 3 p 1657–1666. ISSN/ISBN: 9781510822627.

41. Larive C., Rogat D., Chamoley D., Regnard A., Pannetier T., Thuaud C. (2017) Mid-term Behaviour of Fibre Reinforced Sprayed Concrete Submitted To Flexural Loading. In: Serna P., Llano-Torre A., Cavalaro S. (eds) Creep Behaviour in Cracked Sections of Fibre Reinforced Concrete. RILEM Bookseries, vol 14. Springer, Dordrecht.

42. PD Nieuwoudt, AJ Babafemi, WP Boshoff, 2017, The response of cracked steel fibre reinforced concrete under various sustained stress levels on both the macro and single fibre level, Construction and Building Materials, Vol 156, pp 828–843.

43. Bernard, E.S., 2010. "Influence of Fiber Type on Creep Deformation of Cracked Fiber-Reinforced Shotcrete Panels", *ACI Journal of Materials*, Vol. 107, No. 5, pp474–480.

44. Bernard, E.S., 2012. "The Influence of Creep on Relative Creep Deformations in Shotcrete Linings", *Shotcrete*, Fall, pp. 52–57.

45. Bernard, E.S., 2014. "Creep Rupture of Steel Fibre Reinforced Shotcrete Loaded in Flexure:" Seventh International Symposium on Sprayed Concrete, Modern Use of Wet Mix Sprayed Concrete for Underground Support, 16–19 June, Sandefjord, Norway.

46. WTC Bangkok 2012, "Long-term behaviour of synthetic fibre-reinforced shotcrete in underground construction", Reinhold, M.; Wetzig, V.; Kaufmann, J.

47. Volker Wetzig, Matthias Reinhold, Michael Hermann, Josef Kaufmann, "Long-term behaviour of plastic fibre reinforced sprayed concrete for tunnels", Forschungsprojekt FGU 2010/005_OBF auf Antrag der Arbeitsgruppe Tunnelforschung (AGT), Dezember 2015, No. 1546.

48. Llano-Torre, A, Garcia-Taengua, E, Marti-Vargas, JR and Serna, P, "Compilation and study of a database of tests and results on flexural creep behaviour of fibre reinforced concrete specimens", Proceedings of the FIB Symposium Concrete Innovation and Design, Copenhagen, 2015.

49. P. Serna, A. Llano-Torre, E. García-Taengua and J.R. Martí-Vargas, "Database on the Long-Term Behaviour of FRC: A Useful Tool to Achieve Overall Conclusions", Proceedings of the 10th International Conference on Mechanics and Physics of Creep, Shrinkage, and Durability of Concrete and Concrete Structures, Vienna, September 2015, pp. 1544–1553.

50. ASTM International. C1609/C1609M-19a Standard Test Method for Flexural Performance of Fiber-Reinforced Concrete (Using Beam With Third-Point Loading). West Conshohocken, PA; ASTM International, 2019. https://doi.org/10.1520/C1609_C1609M-19A.

51. Bernard, E.S., 2014. "Influence of Friction in Supporting Rollers on the Apparent Flexural Performance of Third-Point Loaded Fiber Reinforced Concrete Beams," Advances in Civil Engineering Materials, Vol. 3, No. 1, pp. 1–12.

52. Zollo, R.F., 2013. "Analysis of Support Apparatus for Flexural Load-Deflection Testing: Minimizing Bias Caused by Arching Forces," Journal of Testing and Evaluation, Vol. 41, No. 1, pp. 63–68.

53. ASTM International. C1812/C1812M-15e1 Standard Practice for Design of Journal Bearing Supports to be Used in Fiber Reinforced Concrete Beam Tests. West Conshohocken, PA; ASTM International, 2015. https://doi.org/10.1520/C1812_C1812M-15E01.

54. Bernard, E.S., 2019. "Effect of Friction on Performance of Fiber Reinforced Concrete in the ASTM C1550 Panel Test", Advanced Civil Engineering Materials, Vol. 8, No. 1, pp. 285–297.

55. Bjøntegaard, Ø. and Myren, S.A., 2011."Fibre Reinforced Sprayed Concrete Panel Tests: Main Results from a Methodology Study Performed by the Norwegian Sprayed Concrete Committee," in Sixth International Symposium on Sprayed Concrete, Norwegian Concrete Association, Oslo, Norway.

56. Norwegian Concrete Association, 2011. Sprayed Concrete for Rock Support, Report Number 7, p103, Oslo, Norway.

57. Wille, K. and Parra-Montesinos, G.J., 2012. "Effect of Beam Size, Casting Method, and Support Conditions on Flexural Behavior of Ultra-High-Performance Fiber-Reinforced Concrete", ACI Journal of Materials, Vol. 109, No. 3, pp. 379–388.

58. Foster, S.J., Htut, T., and NG, T.S., 2013. "High performance fibre reinforced concrete: fundamental behaviour and modelling", Eighth International Conference on Fracture Mechanics of Concrete and Concrete Structures, FraMCoS-8, (Ed. Van Mier, J., et al), Toledo, Spain.

59. Bernard, E.S., 2015b. "Age-dependent Changes in Post-crack Performance of Fibre Reinforced Shotcrete Linings", Tunnelling and Underground Space Technology, Vol. 49, pp. 241–248.

60. Bjøntegaard, O., Myren, S.A., Klemtsrud, K., and Kompen, R., 2014. "Fibre Reinforced Sprayed Concrete (FRSC): Energy Absorption Capacity from 2 days age to One Year", Seventh International Symposium on Sprayed Concrete, Sandefjord, Norway, 16–19 June, pp 88–97.

61. Bernard, E. S. (2020). Changes in long-term performance of fibre reinforced shotcrete due to corrosion and embrittlement. Tunnelling and Underground Space Technology, 98, 103335. https://doi.org/10.1016/j.tust.2020.103335

62. Kaufmann, J.P., 2014. "Durability performance of fiber reinforced shotcrete in aggressive environment", World Tunnelling Congress 2014, (Ed. Negro, Cecilio and Bilfinger), Iguassu Falls Brazil, p. 279.

Printed in the United States
by Baker & Taylor Publisher Services